Insects

In the same series

Mushrooms and Fungi
Birds
Wild Flowers
Aquarium Fishes
Minerals and Rocks

Chatto Nature Guides

British and European Insects

Identified and illustrated with colour photographs

Wolfgang Dierl

Translated and edited by
Gwynne Vevers

Chatto & Windus · London

Published by
Chatto & Windus Ltd.
40 William IV Street
London WC2N 4DF

*

Clarke, Irwin & Co Ltd
Toronto

British Library Cataloguing in Publication Data

Dierl, Wolfgang
British and European insects.—
(Chatto nature guides).
1. Insects—Europe—Identification
I. Title II. Vevers, Gwynne
595.7'094 QL482.A1

ISBN 0-7011-2380-X
ISBN 0-7011-2379-6 Pbk.

Printed in Italy

Introduction

Any interested observer who comes across insects in the country or even in built-up areas will be faced with the problem of identifying them. There is no easy answer to this as the number of different kinds of insect is quite overwhelming. In Europe alone there are approximately 30,000 different insect species. The majority of these are very small and can only be identified by experts who have access to museum collections and specialist literature. For the layman there are, therefore, considerable difficulties in identifying the insects that he comes across. If he finds a weevil he will be interested primarily in knowing that it is a weevil, and not in trying to find out exactly which species it belongs to. The purpose of this book is to illustrate and describe typical representatives of the more important insect groups found in Britain and Europe. The species selected are those which are common and characteristic. There are, therefore, more species from the better known groups, such as butterflies and moths, than from those such as caddisflies, of which only one representative is described.

Photographs give an even more life-like impression of the insects than good drawings, which in most cases have been made from prepared museum material, often preserved in positions never seen in the wild. Photographs also show a small part of the insect's environment. However, a photograph alone cannot show all the relevant characteristics and so a descriptive text has been prepared which together with the illustrations will aid the task of identification. The book also has a key which describes the special characteristics of the large groups or orders of insects.

The arrangement of the species follows the currently accepted system, i.e. it starts with the more primitive and finishes with the more advanced. The orders and species are given their scientific names. Popular English names are given where they exist, but their use often leads to confusion.

The system of scientific nomenclature, established by Linnaeus, consists of a double name for each species. This consists of a generic and a specific name. Closely related species have the same generic name, that is, they belong to the same genus (plural genera). Related genera are grouped into families, with a name ending in -idae. Related families are placed in orders and, finally, all the orders together make up the class Insecta.
In addition to the scientific name and popular name (where applicable) the descriptive text gives the order to which the species illustrated belongs and also short notes on characteristics, occurrence, diet, life history and habits. In addition, mention is often made of any similar species.

Insect structure

The body of an insect is made up of a number of quite distinct parts. In the adult these are the head, the thorax and the abdomen, but they are not always distinguishable in the larvae.
The head carries the jointed antennae, which are often long, and carry sense organs for touch and smell (fig. p. 8). The head also has the usually large compound eyes, and also some small ocelli, which in primitive insects are usually the only eyes present. Furthermore, the head carries several pairs of mouthparts, which are very important in distinguishing the insect orders (fig. p. 8). The most primitive mouthparts are those adapted for biting, with two mandibles which work against one another. In insects which feed on liquid food the mouthparts are modified to form a tube for sucking. Finally, there are insects which in addition to sucking also pierce their food, whether animal or plant. In many cases the larvae have mouthparts differing in structure from those of their adults.
The thorax consists of three parts (prothorax, mesothorax and metathorax), each of which carries a pair of legs on the ventral side. The legs are jointed, with coxa, femur, tibia and foot or tarsus. In some insects the larval legs differ in form from those of the adult. The legs may also be furnished with various appendages, such as spines and claws. In most insects the mesothorax and metathorax each carry a pair of wings, which consist of a thin membrane strengthened and held taut by veins; the arrangement of the veins is very important in distinguishing the orders. The wings are often coloured and may carry hairs or sceles. In the primitive insects there are no wings and the larvae of insects with complete metamorphosis

(see below) are also wingless. On the other hand, in the larvae (often called nymphs) of insects with incomplete metamorphosis the wing rudiments are already visible externally as small stumps.

The front and hind wings may differ considerably in form as, for example, in the beetles where the front wings are modified as wing-covers or elytra, while the hind wings are larger and remain membranous. In the flies the hind wings are reduced to small club-shaped halteres. In the course of evolution the wings of some insects have become atrophied and such insects are then flightless. The position of the wings when at rest also varies. They may be extended backwards over the body as in crickets and cockroaches, or folded away under hard wing-covers as in beetles and earwigs. In the butterflies and moths they are either extended outwards or held together over the body.

The abdomen consists of a series of similarly shaped segments, which together form a tube. It is either soft and is then covered by hard wing-covers or it is enclosed by hard dorsal and ventral plates. It may be hairy or scaly. In the less advanced insects the rear end of the abdomen may carry two or three caudal appendages, in female grasshoppers an ovipositor (egg-laying apparatus) or in earwigs a pair of pincers. In wasps, bees and related insects the ovipositor is modified to form a sting associated with a venom gland. In other insects the ovipositer is hidden within the body. This also applies to the male's reproductive organs.

The internal organs comprise the alimentary tract, the nervous system (as ventrally situated nerves), a blood system and numerous muscles. In addition there are also thin tubes, the tracheae, which run from lateral openings, the stigmata, on the thorax and abdomen. The tracheae carry air for respiration to the internal organs. Finally there are the reproductive organs with openings at the rear end of the body. The outer covering of the insects is strengthened on all or some parts of the body by a hard cuticle which protects the body, gives it a shape and as an external skeleton (exoskeleton) provides a fixture for the musculature. This strengthened skin or cuticle consists of chitin, a carbohydrate similar to cellulose, and sclerotin, a hardened protein. These two substances together form a very hard and resistant, but very light structure which in the course of evolution has served a variety of functions. Finally, the surface of the cuticle is

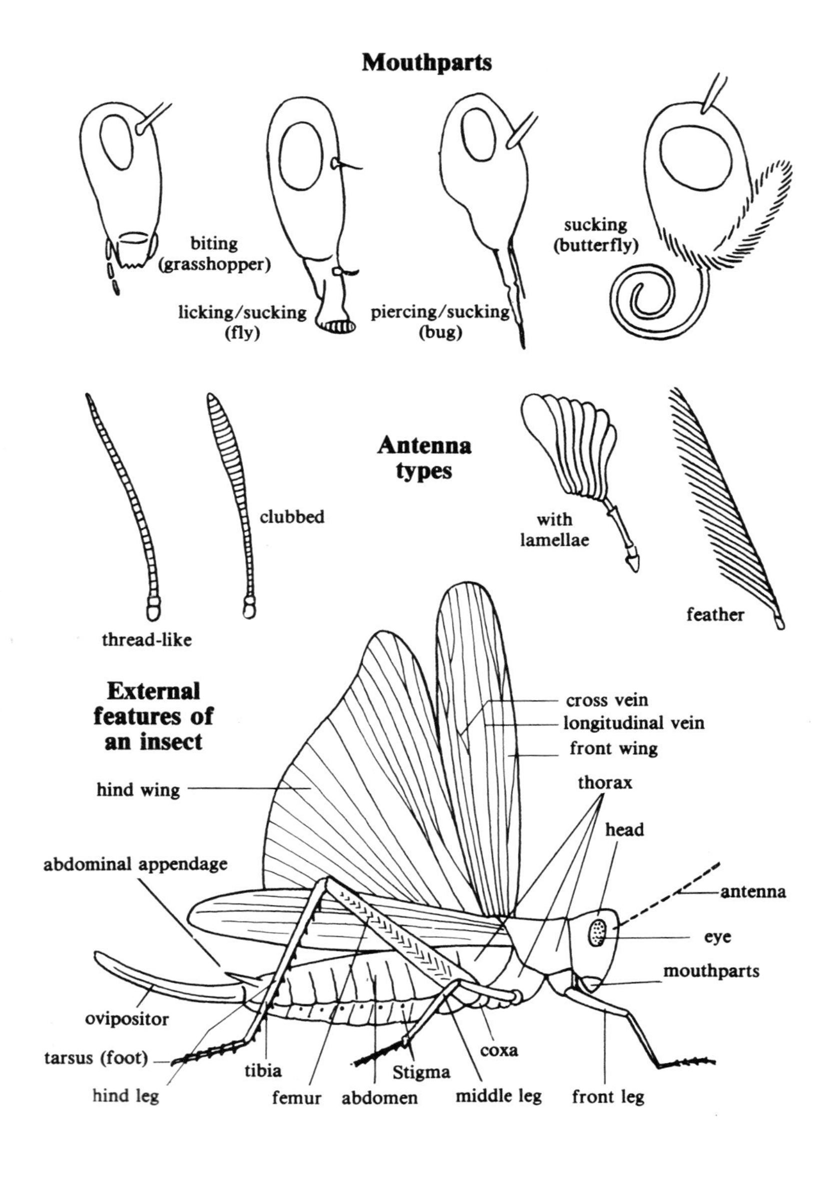
Mouthparts
biting
(grasshopper)
licking/sucking
(fly)
piercing/sucking
(bug)
sucking
(butterfly)
Antenna
types
clubbed
thread-like
with
lamellae
feather
External
features of
an insect
cross vein
longitudinal vein
front wing
hind wing
thorax
head
abdominal appendage
antenna
eye
mouthparts
ovipositor
tarsus (foot)
tibia
Stigma
coxa
hind leg
femur
abdomen
middle leg
front leg

covered by a waxy layer which acts as a water repellant. This very resistant and useful form of skin has, however, one great disadvantage. Once formed it cannot be altered, that is, it can no longer grow. How this problem is solved in the course of an insect's development will be described in the following section.

The development of insects

The life history of an insect, like that of other animals, starts with the egg. Insect eggs differ considerably in form and coloration, and they are enclosed in a chitinous casing. They are usually very small and difficult to find. A few examples are illustrated. The larva that hatches from the egg is naturally also very small, and it starts to feed and grow. Soon, however, it reaches the capacity of its non-growing skin. It then undergoes a moult, in which it throws off the old skin and a new larger skin appears which allows growth to continue. This process is repeated several times until the larva has reached its full size. The number of larval stages, between the moults, is quite specific. This means that almost every species has a well-defined number of moults.
There are two types of insect development. These are known as incomplete and complete metamorphosis. An insect with incomplete metamorphosis develops continuously, so that at each moult the young stage (nymph) becomes more like the adult. The wing rudiments appear quite soon as small stumps and become larger at each moult. The mouthparts of the nymph are very similar to those of the adult insect and function in the same way, so that the diet of the two stages is very similar. In an insect with complete metamorphosis, on the other hand, the process is quite different. Here the larva changes very little during its development, except that it becomes larger. When it has reached a certain size it changes into the pupal stage which differs in appearance from the larva, does not feed and scarcely moves. This happens, for instance, in butterflies and moths (see illustration p. 10 of book). In the pupa the larval organs are broken down and the resulting material, together with substances stored by the larva, are used in the formation of the adult insect. When this process is complete the pupal skin is shed and the adult insect (or imago) emerges. At this moment the wings are still small, soft and much folded. Blood pumped through the wing veins stretches the wing membranes to their full size and they then

Larval types

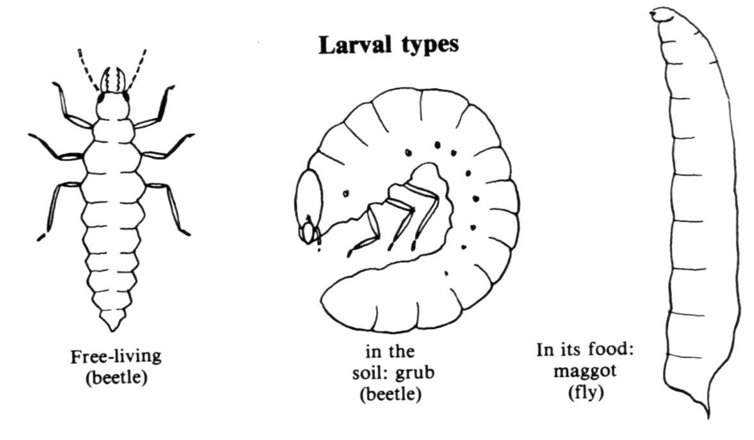

Metamorphosis

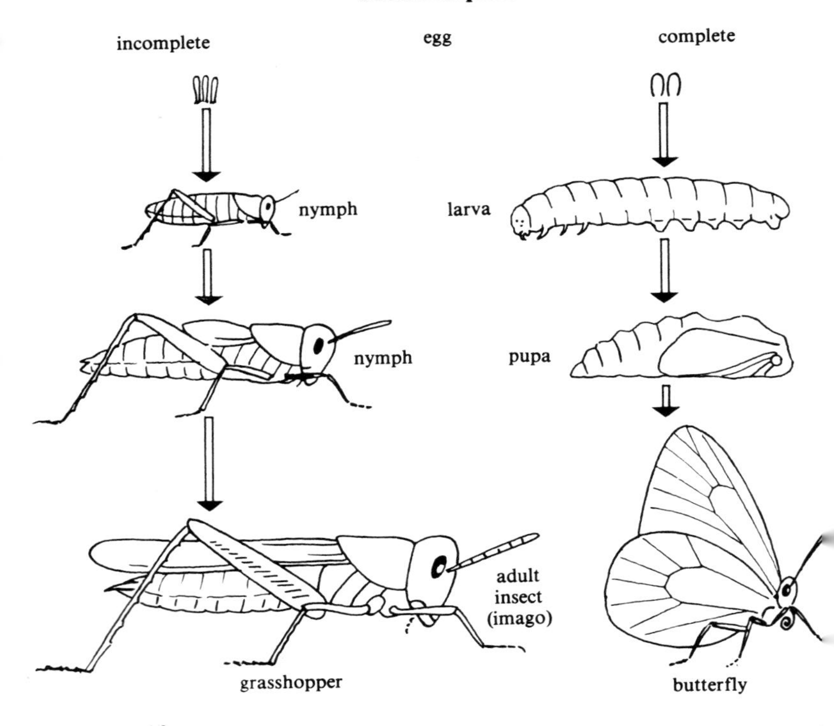

harden in the air. Other parts of the body that eventually become hard undergo the same process. The term complete metamorphosis refers to the great difference between the larva, pupa and adult. These differ not only in general appearance but also in the structure of the mouthparts. Thus, in butterflies and moths, for example, the larvae feed on leaves whereas the adults suck nectar.

Finally, in the really primitive insects development is direct, without any special larval form.

In temperate climates insect development is necessarily interrupted during the winter, whereas in the tropics it proceeds continuously. All living processes are dependent upon temperature, and so in winter in temperate climates there is a resting period during which almost all bodily functions are at a standstill. In theory any developmental stage of an insect could overwinter, but usually each species only overwinters in one particular stage. In the simplest case the existence of such overwintering stages is brought about by the falling temperatures in autumn. In higher insects, on the other hand, the decreasing day length in autumn stimulates the onset of the resting stage.

An insect species may appear once or several times in the year, depending upon its rate of development during the summer. One then speaks of one or more broods. In certain cases these may look quite different, as can be seen quite clearly in the Map Butterfly on p. 71. Here the different colours are dependent primarily upon the different temperatures during pupal development.

Insect larvae vary considerably in general appearance (fig. p. 10), and so it is not surprising that some have been given special names. The differences in larval structure are correlated with their habits and are reflected in the systematic position of the insects concerned. Aquatic larvae are different in appearance from those that live on land, and have a different method of respiration. Similarly, vegetarian larvae differ from those that are predatory. This is seen very clearly among the beetles which show an incredible range of larval form and general habits. Here the best known larvae are those with three pairs of legs and a thick, curved body which is not very mobile; these are often referred to as grubs. The larvae of butterflies and moths, known colloquially as caterpillars, have three pairs of legs on the thorax and five pairs or even fewer on the abdomen. These are vegetarian. Finally, there are legless larvae, often known as maggots. These occur, for instance, among the flies, in which the larva may live in or on its food.

Finally, some mention must be made of the phenomenon of parasitism. There are certain insects and in particular their larvae which live and develop in or on other animals. These include, for instance, the numerous parasitic ichneumons and other hymenopterans which spend their larval life inside the body of other larvae, e.g. of butterflies and beetles, on which they feed. This, of course, leads to the death of the host larva. The adult ichneumon pierces the skin of the host with its ovipositor and lays an egg inside the host's body. Other parasitic hymenopterans paralyse their prey by stinging and lay an egg on the outside of its body.

Identification key for insect orders

	Wingless, with caudal appendages or a forked springing organ; very small Order Apterygota	**p. 16**
	(There are also other orders with wingless species, but they never have long caudal appendages or a forked springing organ)	
1a	Winged insects	**2**
2	With 1 pair of wings, the hind wings reduced to small knobs; no caudal appendages (drawing No. 1) Order Diptera	**p. 20 and pp. 98-104**
2a	With 2 pairs of wings	**3**
3	Both pairs of wings membranous, with veins	**4**
3a	Front wings hard or leathery, opaque	**12**
4	Wings covered with coloured scales; usually with a long proboscis Order Lepidoptera	**p. 19 and pp. 60-96** (drawing No. 2)
4a	Wings hairy or naked, usually transparent	**5**
5	Longitudinal wing veins connected by numerous cross veins	**6**
5a	Only a few cross veins present	**9**
6	Abdomen with long caudal appendages	**7**
6a	Caudal appendages very short or absent	**8**
7	Front wings triangular and much larger than hind	

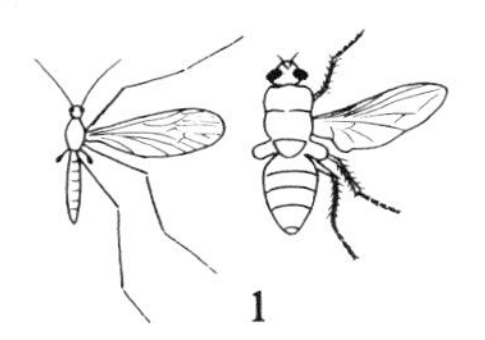

1

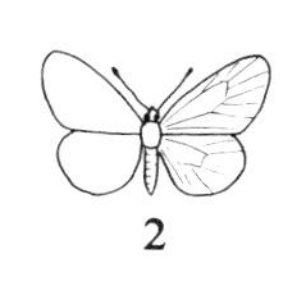

2

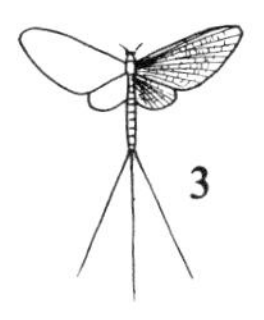

3

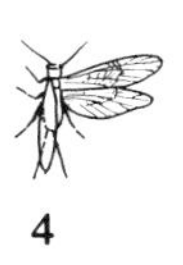

4

wings; at rest, the wings are held vertically over the body; 2-3 caudal appendages

Order Ephemeroptera **p. 16 and p. 22**
(drawing No. 3)

7a Front and hind wings equal in size, longish, and in rest held flat on the back; 2 caudal appendages
(drawing No. 4)

Order Plecoptera **p. 17 and p. 32**

8 Antennae very short, less than the width of head; body long, brightly coloured

Order Odonata **p. 16 and pp. 24-32**
(drawing No. 5)

8a Antennae much longer, several times the width of the head

Order Neuroptera **p. 19 and pp. 52-58**
(drawing No. 6)

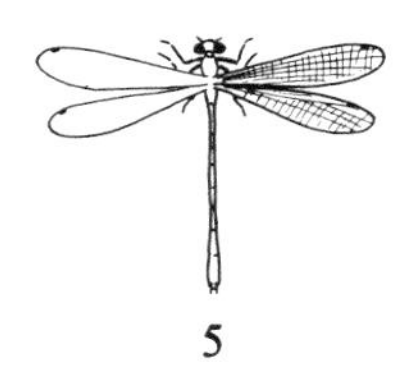

5

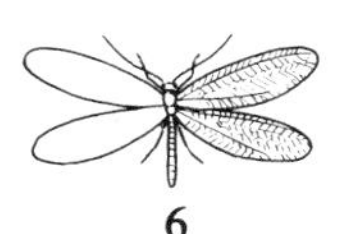

6

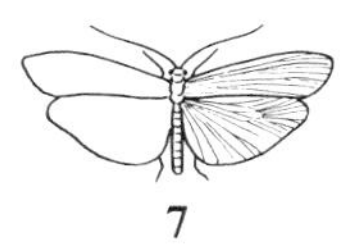

7

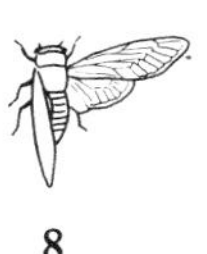

8

9 Wings hairy, hind wings larger than front wings

Order Trichoptera **p. 20 and p. 58**
(drawing No. 7)

9a Wings naked **10**

10 Large insects with broad head, short antennae and piercing and sucking mouthparts

Order Hemiptera **p. 18 and pp. 44-52**
(drawing No. 8)

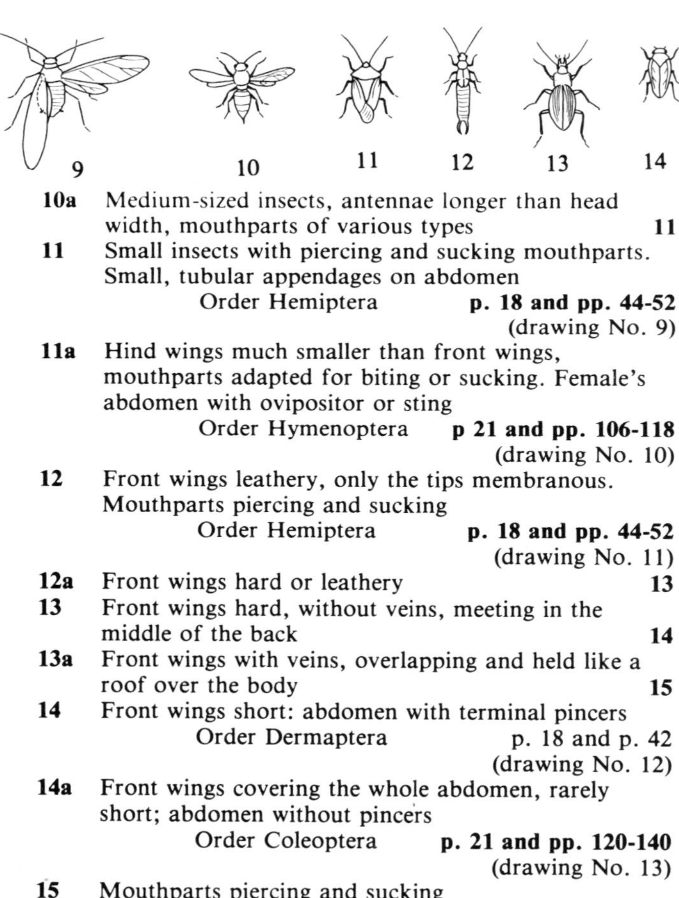

10a Medium-sized insects, antennae longer than head width, mouthparts of various types **11**

11 Small insects with piercing and sucking mouthparts. Small, tubular appendages on abdomen
Order Hemiptera **p. 18 and pp. 44-52**
(drawing No. 9)

11a Hind wings much smaller than front wings, mouthparts adapted for biting or sucking. Female's abdomen with ovipositor or sting
Order Hymenoptera **p 21 and pp. 106-118**
(drawing No. 10)

12 Front wings leathery, only the tips membranous. Mouthparts piercing and sucking
Order Hemiptera **p. 18 and pp. 44-52**
(drawing No. 11)

12a Front wings hard or leathery **13**

13 Front wings hard, without veins, meeting in the middle of the back **14**

13a Front wings with veins, overlapping and held like a roof over the body **15**

14 Front wings short: abdomen with terminal pincers
Order Dermaptera p. 18 and p. 42
(drawing No. 12)

14a Front wings covering the whole abdomen, rarely short; abdomen without pincers
Order Coleoptera **p. 21 and pp. 120-140**
(drawing No. 13)

15 Mouthparts piercing and sucking
Order Hemiptera **p. 18 and pp. 44-52**
(drawing No. 14)

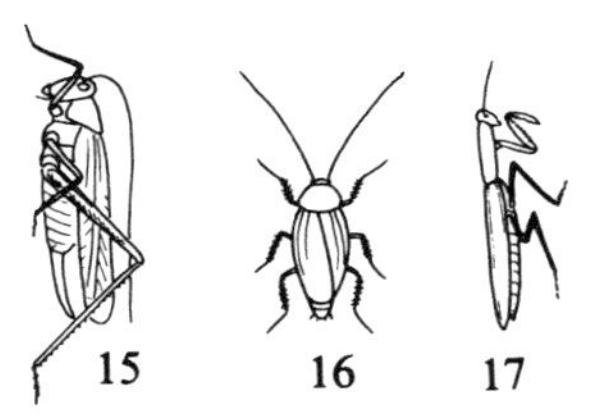

15a	Mouthparts adapted for biting	**16**
16	With long legs for jumping Order Saltatoria	**p. 17 and pp. 34-38** (drawing No. 15)
16a	Walking legs not adapted for jumping Order Blattodea	**p. 18 and p. 42** (drawing No. 16)
16b	Front legs modified for seizing Order Mantodea	**p. 17 and p. 40** (drawing No. 17)

Description of the major groups in the Class Insecta

Sub-class Apterygota

These are small or very small, primitive insects which at no time in their evolutionary history have had wings. They have biting mouthparts and long caudal appendages or a forked springing organ at the rear end of the body with which they can jump into the air. Development is direct and the young can scarcely be distinguished from the adults, as they have no special larval characteristics. These are inconspicuous insects which live on or in the soil, sometimes in enormous numbers. They are not illustrated here.

Sub-class Pterygota

These are insects with wings, which in certain cases may have been lost during the course of evolution, e.g. in fleas. Metamorphosis is incomplete or complete. All the orders dealt with in this book belong here.

Order Ephemeroptera (mayflies) Page 22

Distinguished from similar groups by the very short antennae. Front wings large, triangular, hind wings small or absent. Abdomen with 2-3 long caudal appendages. Delicately built insects, without conspicuous colours, and usually found near water in which the larvae develop. The last larval (or nymphal) stage is capable of flying; it then moults to produce the adult mayflies. About 70 species in Europe.

Order Odonata (dragonflies and damselflies) Pages 24-32

Mostly large, brightly coloured insects with two pairs of wings with a network of veins, large eyes, very short antennae and a long, slender abdomen. They are predatory and catch their prey in flight. The nymphs are aquatic and undergo an incomplete metamorphosis. They too are predatory, feeding on various invertebrates and fish fry. Their mouthparts are modified to form a mask adapted for seizing prey. There are two sub-orders, the Zygoptera or damselflies (pp. 24, 26, 32) with the wings all similar in shape and the Anisoptera or dragonflies (pp. 28-30) with broader hind wings. In the Zygoptera the wings are held over the back when at rest, but in the Anisoptera they are extended sideways. About 80 species in Europe.

Order Plecoptera (stoneflies) Page 32

Medium-sized, inconspicuous insects with membranous, almost equal-sized wings, which are laid flat over the body when at rest. Antennae long and thin. Mouthparts usually reduced. The abdomen usually has two caudal appendages, thus distinguishing stoneflies from the similar caddisflies and lacewings. Stoneflies are always found in the vicinity of water, particularly running water, in which their nymphs live and develop. They undergo an incomplete metamorphosis, feed mainly on plant matter and are themselves an important food for fishes. As they are very sensitive they are good indicators of the purity of the water. About 112 species in Europe.

Order Saltatoria (grasshoppers and crickets) Pages 34-38

These are medium-sized or largish insects, often attractively coloured, with biting mouthparts and rather large eyes. Prothorax enlarged and saddle-shaped. The hind wings are broader than the front wings and sometimes brightly coloured. Hind legs long and powerful and used for jumping. These insects can produce sounds. Metamorphosis is incomplete. About 130 species in Europe, in three main families:

Family Gryllidae (crickets) (p. 34).

Broad, flattened saltatorians with short front wings, and long caudal appendages. Sounds are produced by rubbing the wings against one another.

Family Tettigoniidae (bush crickets) (pp. 34-38).

Saltatorians with very long, thin antennae, living particularly among bushes, and often active at night. Sounds produced by rubbing the wings against one another.

Family Acrididae (grasshoppers and locusts) (p. 36).
Saltatorians with short antennae, living on the ground and mainly active during the day. Sounds are produced by rubbing pegs on the hind legs against a hard vein on the front wings.

Order Mantodea (mantises) Page 40
Similar to grasshoppers but with the hind legs not adapted for jumping. Front legs modified for seizing prey. Head very mobile, with biting mouthparts. Metamorphosis is incomplete. Mantises feed primarily on other insects. The order is represented mainly in the tropics. They are often grouped with the cockroaches to form the order Dictyoptera. Very few species in the warmer parts of Europe.

Order Blattodea (cockroaches) Page 42
Medium-sized, flattened insects with the hard front wings lying flat, one above the other. Antennae long, mouthparts biting and abdomen with caudal appendages. The legs are similar in form and adapted for running. Metamorphosis is incomplete. Cockroaches live on the ground and are mainly active at night. They and their larvae are omnivorous. The group is mainly distributed in the tropics, but there are about 16 species in Europe, some of which have been introduced from warmer parts of the world.

Order Dermaptera (earwigs) Page 42
Elongated, brownish insects with short front wings (elytra) and powerful pincers at the rear end of the abdomen. Earwigs have biting mouthparts and feed mostly on plant matter, but also take some animal food. They are mainly nocturnal and use the relatively large hind wings for flying. Metamorphosis is incomplete and the females practice a form of brood protection, which is astonishing for such primitive insects. About 7 species in Europe.

Order Hemiptera (true bugs) Pages 44-52
Small to large insects, always with piercing and sucking mouthparts, but otherwise very variable in appearance and habits. The majority suck plant juices, a smaller number live as predators. Metamorphosis is incomplete. About 2400 species in Europe. The order is subdivided into two groups according to the wing structure:
Sub-order Heteroptera (pp. 44-48).
Front wings leathery, except the tips which are membranous.

Body usually flattened, often brightly coloured and varying in form. Feeding on plants or as predators.
Sub-order Homoptera (pp. 50-52).
Front and hind wings always similar, either transparent with veins in some cicadas and aphids or quite leathery and often brightly coloured in the smaller cicadas. In certain aphid forms the wings are lacking. In contrast to the Heteroptera the bugs in this group are more thick set, the body usually having a circular cross section. Unlike the cicadas the aphids have small tubes on the abdomen which are the openings of wax glands. Homopterans are all purely vegetarian and some cause very serious damage to crops. Large, loud-singing cicadas occur in southern Europe but not so abundantly as they do in the tropics.

Order Neuroptera (lacewings, alderflies, snakeflies)
Pages 52-58
Insects of various sizes with long antennae and membranous wings having numerous longitudinal and transverse veins. At rest, the wings are folded rooflike over the body. Distinguished from similar groups by the absence of caudal appendages, the long antennae and the non-hairy wings. Metamorphosis is complete. There are two sub-orders:
Sub-order Megaloptera (alderflies and snakeflies) (pp. 52-54).
Alderflies are largish, dark insects, always living in the vicinity of water, in which the larvae develop. Distinguished from the other Neuroptera by the absence of forked veins at the wing edges. Three species in Europe.
Snakeflies are smaller with an elongated pre-thorax forming a long "neck", and biting mouthparts. The larvae are terrestrial predators living in bark or tunnels in timber. Twelve species in Europe.
Sub-order Planipennia (lacewings) (pp. 54-58).
The adults differ considerably in form and coloration, but all have biting mouthparts and heavily veined wings. The larvae differ from those of other Neuroptera in having sucking mouthparts with which they suck the body contents of their prey. Some larvae are aquatic, others have evolved special capture techniques on land. About 100 species in Europe.

Order Lepidoptera (butterflies and moths) Pages 60-96
Well-known insects with membranous wings covered with brightly coloured scales. In some species the scales may be absent from parts of the wings, thus producing transparent

"windows". The mouthparts are in the form of a proboscis which can be coiled up, and which serves for sucking up plant juices. There are some species, however, in which the proboscis is reduced. Some inconspicuous small species, the Micropterigidae, have functional mandibles and feed on pollen. The antennae are always well developed and they vary in form depending upon the systematic position and habits. In nocturnal moths they are frequently thickened, serrated or feathery. Antennae carry sense organs which help the insect to find food and a mate. Diurnal species usually have simple antennae, with a club-like thickening at the end. Metamorphosis is complete, with egg, larva (here known as a caterpillar), pupa and imago or adult. The caterpillars differ considerably in appearance, and may be naked, hairy or spiny, and many are brightly coloured. The pupae hang free, or lie in a cocoon or a hole in the ground. More than 3000 species in Europe.

Order Trichoptera (caddisflies) Page 58

Mostly medium-sized insects with membranous wings that only have a few veins. The wings are hairy. Distinguished from other orders by the hairy wings and the small number of veins. The biting mouthparts are poorly developed. Caddisflies are inconspicuously coloured and usually live in the vicinity of water where they are active in the evening or at night. The larvae are mostly aquatic, living in very characteristic self-made cases, the construction of which varies according to the species. The larval feeding habits are very variable and some are predatory. The larvae form an important food for fishes. About 300 species in Europe.

Order Diptera (true flies) Pages 98-104

a very characteristic group in which the hind wings are reduced to small club-like halteres or balancers. The mouthparts are modified for sucking or piercing and sucking. The Diptera are divided into three sub-orders, distinguished primarily by the form of the antennae. About 8000 species in Europe.

Suborder Nematocera (pp. 98-100).

Characterised by the long, thin antennae, slender body and usually long, thin legs. The habits of the larvae are very variable; some are aquatic, some terrestrial, others live in or on their food, while some are responsible for gall formation. The adult feeding habits range from sucking plant juices to

sucking blood, while those with reduced mouthparts do not feed at all.
Sub-orders Brachycera and Cyclorrhapha (pp. 100-104).
Both groups are rather similar to one another, and have the typical fly shape. They are more thick-set than the Nematocera and have much shorter antennae. They are also more brightly coloured and often show brilliant iridescence. The mouthparts are modified for sucking or piercing and sucking. The larvae are characterised by the reduction of the head and legs and are often known colloquially as maggots.

Order Hymenoptera (ants, bees and wasps) Pages 106-118
Insects of very variable size with two pairs of membranous wings which have a few veins. The mouthparts are modified for biting or sucking, and metamorphosis is complete. They are classified in two major groups, distinguished by the presence or absence of a "wasp waist". Their habits vary considerably, and there are highly specialised social and parasitic forms. About 10,000 species in Europe.
Sub-order Symphyta (p. 106).
Hymenopterans characterised by the absence of a "wasp waist". They feed mainly on plants. Larvae with at least 6 pairs of abdominal legs.
Sub-order Apocrita (pp. 106-118).
Hymenopterans with a well-defined "wasp waist", often living socially or as parasites. In many the ovipositor is modified to form a venomous sting. Some species are wingless. The diet varies from plants and their juices to live prey. The larvae are maggot-like with reduced head and legs.

Order Coleoptera (beetles) Pages 120-140
Insects of very variable size with two pairs of wings. The front wings, known as the elytra, are always hard or leathery, and they never overlap. The membranous hind wings lie folded beneath the elytra. The mouthparts are modified for biting and metamorphosis is complete. The habits and diet of both adults and larvae vary considerably. Beetles have colonised almost every part of the world. They are distinguished from other similar insects (e.g. cockroaches) by the absence of caudal appendages. About 6000 species in Europe.

Insects

Mayfly *Ephemera vulgata*

Order Ephemeroptera (mayflies)

Characteristics: a delicately built insect with large, triangular membranous front wings with numerous veins, and long thin "tails" or caudal appendages. When at rest the wings are folded together over the back. The mouthparts of the adult insects are reduced. The aquatic nymphs also have long "tails" and they carry tracheal gills on the sides of the abdomen with which they take up oxygen from the water.—**Occurrence:** always in the vicinity of water, which is usually clear and running.—**Diet:** the nymphs feed mainly on algae and plant debris, but the adults do not feed.—**Life history and habits:** the nymphs undergo a series of moults over a period of 1-2 years, to form a sub-imago which resembles the adult or imago and is capable of flight. The sub-imago which is now resting on a plant stem then moults into the adult. The adults swarm in large numbers, particularly in the evening, mate and the females lay their eggs. Adults only live for a few hours or days, hence the Ephemeroptera (Greek *ephemeros,* living for a day). The large numbers of mayflies provide welcome food for birds and bats and dead adults in the water are quickly consumed by fishes. Huge numbers of these swarming insects can occasionally darken the sky.

There are relatively few mayfly species, which occupy differing ecological niches and are not easy to identify. Characters used in identification include the coloration, the wing venation and the number of "tails". The nymphs are easier to identify because in adapting to certain environmental factors they have evolved characteristic forms.

White-legged Damselfly *Platycnemis pennipes*

Order Odonata (dragonflies and damselflies)

Characteristics: body very slender, thin and blue with black markings. Wings completely transparent with only a small spot close to the tip. Body length 35 mm. Antennae very short, eyes very large. Legs with feathery processes. The male is more brightly coloured than the female. The illustration shows a pair mating.—**Occurrence:** in summer close to quiet stretches of water, often over reed beds, where these insects rest at night.—**Diet:** both nymphs and adults are predators that seize their prey with powerful jaws. The adults feed on small flying insects, including mosquitoes and may therefore be very beneficial. The nymphs, which live and feed in the water, have three narrow, leaf-like appendages on the abdomen (tracheal gills), which distinguish them from other aquatic insect nymphs.—**Life history and habits:** before mating the male fills special reproductive organs on his 2nd and 3rd abdominal segments with sperm. He does this by bending the body so that the genital opening at the end of his body is brought up to these special organs. He then finds a female and grips her behind the head with the "claspers" on his 10th abdominal segment. The female now bends her body forwards until her genital opening meets the special reproductive organs of the male where she collects the sperm. The female is still held in this position during egg-laying, and both insects fly around in tandem. The eggs are laid on the stem of water plants, usually underwater. The tandem position may help to protect the female during egg-laying. The eggs hatch in the water and the nymphs develop there, taking dissolved oxygen through the tracheal gills at the hind end. After several moults over a period of less than a year the nymphs are fully grown and they then climb out of the water to emerge as adult damselflies.

European damselflies are very similar to one another, but the species can be distinguished by coloration and pattern, the arrangement of the wing veins and the structure of the legs and tracheal gills.

Variable Coenagrion *Coenagrion pulchellum*

Order Odonata (Dragonflies and damselflies)

Characteristics: this is a characteristic damselfly with a slender body and transparent wings. Body length 30 mm. The coloration is iridescent blue with black markings. Many males have a U-shaped mark on the second segment of the abdomen, but this mark is more thistle-shaped in the female. Distinguished from closely related species mainly by the form of the markings. In the females there are two colour variants, one bluish-black (like the males) and the other greenish-black. As in all members of the order the eyes of damselflies are very large and consist of numerous separate visual units. In damselflies and dragonflies there may be up to 30,000 facets in each eye. The size of the eyes is important as these insects catch their prey on the wing and thus require excellent vision. The legs which are furnished with powerful claws and spines, are held forwards during flight and together form a "basket" in which the prey is caught. The mouthparts are adapted for biting and consist of powerful, toothed mandibles with which the prey is chewed.—**Occurrence:** widely distributed in the vicinity of quiet waters. The adults are seen throughout the summer and when the sun is out they can often be seen hunting, mating and egg-laying.—**Diet:** this is described on p. 24.—**Life history and habits:** the nymphs undergo several moults over a period of a year, and then leave the water and moult into the adult insect. It takes a few days before the full adult coloration is attained, young adults being very pale. Damselflies are only active by day, particularly in sunlight, and spend the night among reeds and sedges, with wings folded and often covered with drops of water. At such times they can be observed at close quarters and even picked up. During mating the male clasps the neck of the female. The eggs are usually laid on the underside of water plants, usually with the male still holding the female, who inserts her eggs into the plants. The nymphs live and develop in the water, hunting for food on the bottom. Their inconspicuous mottled brown coloration provides good camouflage.

Southern Aeshna *Aeshna cyanea*

Order Odonata (dragonflies and damselflies)

Characteristics: a dragonfly with a long thin body, stouter than in the damselflies, with green and yellowish coloration and black markings. The wings are transparent with small dark flecks near the tips. Body length up to 80 mm. The iridescent green or blue eyes are very large and in contact with one another on top of the head. There are a few other related species of the same size, which can be distinguished by coloration and pattern.—**Occurrence:** widely distributed near still waters, in which the nymphs live (opposite, above left). Dragonflies are often seen quite far from water, sometimes in forest clearings where they hunt for prey. The adults are seen from June to October.—**Diet:** the adults and nymphs are greedy predators which can take quite large prey. The adults take advantage of air currents when hunting, and they are excellent fliers, which move with lightning speed and so can even catch fast-moving insects. The wings can move independently of one another, allowing the insect to remain motionless in the air, or even move backwards for short distances.—**Life history and habits:** in accordance with the size of the adults, the nymphs require plenty of food and, depending upon the temperature, they often take several years to complete their development. The final nymphal stage climbs out of the water and moults to produce the adult. While its abdomen is still attached to the nymphal skin the new adult hangs head downwards, completely motionless. It then bends its body upwards, clings with its legs to the larval skin and pulls its abdomen free. At this stage the wings are still very small, but blood is now forced through the veins to inflate them to their full size (opposite, above right). The wings are still soft and it takes several hours in the air before they become hard and the insect can fly away. On the other hand, the attractive colours do not become fully devloped for several days. In spite of the old country belief that dragonflies can sting this is in fact quite untrue and they are completely harmless.

Black-lined Orthetrum *Orthetrum cancellatum*

Order Odonata (dragonflies and damselflies)

Characteristics: wings with only one dark spot near the tip and without dark coloration at the base. The abdomen is blue with a black tip and black base and is not so broad as in the preceding species. Body length c. 40 mm. Here too the female is less conspicuously coloured than the male.—**Occurrence:** not uncommon in the vicinity of quiet waters throughout the region.—**Diet:** flying insects which are caught in flight. The aquatic nymphs live among vegetation on the bottom and catch worms, insect larvae and other prey.—**Life history and habits:** nymphal development may take several years, depending upon the food supply and the water temperature. After its emergence the adult dragonfly starts to fly in the characteristic manner of the group.

Four-Spotted Libellula *Libellula quadrimaculata*

Order Odonata (dragonflies and damselflies)

Characteristics: this dragonfly is characterized by the two conspicuous dark spots on the front edge of each wing. The body is compressed and broader than in other dragonflies, and is blue with yellow markings on the sides in the males, inconspicuous brownish in the females. Body length up to 50 mm.—**Occurrence:** common and widely distributed near quiet waters, and seen in flight from May to August.—**Diet:** all stages are predatory. The adults often lie in wait in reed beds for passing insects.—**Life history and habits:** the nymphs live among vegetation on the bottom and their development takes 2-3 years. The adults have a relatively long life span of several weeks and they move about quite a lot. At night and during bad weather they hide away in vegetation and are then rarely seen.

Demoiselle Agrion *Calopteryx splendens*

Order Odonata (dragonflies and damselflies)

Characteristics: this damselfly, also known as *Agrion splendens,* has conspicuous iridescent colours on the body and wings, the male having a broad blue band in the middle, the female being greenish. As in all the smaller damselflies the body is very thin and scarcely 50 mm long.—**Occurrence:** widely distributed in low-lying country along slow-flowing rivers and canals with tall vegetation on the banks.—**Diet:** the adults hunt flying insects, the nymphs feed on bottom-living invertebrates.—**Life history and habits:** nymphal development usually takes two years and the adults live for about two weeks. In sunshine they fly slowly over the water and often rest on the plants.

Stone Fly *Perlodes* sp.

Order Plecoptera (stoneflies)

Characteristics: an inconspicuously coloured insect with long antennae and two caudal appendages on the abdomen. At rest the wings lie folded flat over the body.—**Occurrence:** these insects mostly live a rather secluded life along the banks of clear, slow-flowing rivers. They were formerly more abundant but are now less frequently seen as their numbers have declined following the pollution of so many waters. They appear mainly in the spring, in some places quite soon after the snow has melted.—**Diet:** the adults have weak mouthparts and scarcely feed at all. The nymphs are aquatic, feeding mainly on algae and other plants but some hunt for prey.—**Life history and habits:** development usually takes one year and the nymphs crawl out on to land before emergence of the adult. All stages of development provide a source of food for fishes, particularly trout. The adults do not fly much, but the females often lay while in flight, by dipping the tip of the abdomen in the water and releasing the eggs.

Field Cricket *Gryllus campestris*

Order Saltatoria (grasshoppers and crickets)

Characteristics: a shiny black insect with a broad, round head and long antennae. The wings are folded flat over the back. The hind legs are modified for jumping, but are not so long as in the grasshoppers. Body length 22 mm.—**Occurrence:** widely distributed in Europe and quite common on grassy slopes and along the edges of fields. The adults can be found and heard during spring.—**Diet:** the nymphs and adults feed mainly on plants.—**Life history and habits:** crickets are nocturnal insects that remain hidden by day in holes which they dig themselves. The nymphs overwinter and the adults emerge in the following spring. The males produce characteristic sounds by rubbing the wings against one another, and these serve to attract the females. The auditory organs lie in the tibiae of the front legs. Field Crickets sing until the end of June, any songs heard later being the work of grasshoppers.

Great Green Bush Cricket *Tettigonia viridissima viridissima*

Order Saltatoria (grasshoppers and crickets)

Characteristics: large green insects with long antennae. The females have a long, sword-like ovipositor. The hind legs are long and adapted for jumping. When at rest the wings are held roof-like over the body and they extend beyond the body which is up to 55 mm long.—**Occurrence:** widely distributed and not uncommon among bushes and trees. The adults are seen from July onwards.—**Diet:** the nymphs and adults feed on animal and plant food.—**Life history and habits:** the eggs overwinter and hatch in spring. Development is completed during the summer, when the song of the males, produced by rubbing the wings, can be heard. The song changes when a female comes close to the male. After mating the female lays her eggs in the ground or in crevices, using the long ovipositor. This species flies for short distances.

Barbitistes serricauda

Order Saltatoria (grasshoppers and crickets)

Characteristics: a bush cricket with long antennae. The females, which are flightless, have a curved ovipositor which is serrated at the top. This enables it to pierce the bark crevices in which the eggs are laid.—**Occurrence:** a warmth-loving species, occurring mainly in the southern part of the region, and not in Britain. Seen in bushes on the edge of woodland, but not very commonly.—**Diet:** the adults and nymphs are predatory, feeding on invertebrates, particularly insects.—**Life history and habits:** the eggs overwinter and hatch in spring. The nymphs undergo several moults and become fully grown in a few months. These bush crickets rarely jump, but climb about among bushes. They are mainly active at night and do not usually start to sing until the evening.

Psophus stridulus

Order Saltatoria (grasshoppers and crickets)

Characteristics: the antennae of this grasshopper are very much shorter than those of the preceding species and so are the wings which scarcely extend beyond the abdomen. The coloration is a spotted grey-blue which provides excellent camouflage when the insect is at rest. The red hind wings are hidden when the cricket is resting, but they give a bright flash when it flies off. When flying the males emit a loud, rattling sound.—**Occurrence:** seen from summer to autumn in warm places, particularly on heathland. Not found in Britain.—**Diet:** the adults and nymphs feed on plants.—**Life history and habits:** development is the same as in other grasshoppers. These insects live mainly on the ground, and when they take off the shock effect of the unexpected red flash is enhanced by the sound emitted by the males. The sounds of grasshoppers are produced by rubbing the hind legs on the wings. They hear with organs situated on the sides of the abdomen. The females can also produce sounds, but these differ from those made by the males.

Wart-biter *Decticus verrucivorus*

Order Saltatoria (grasshoppers and crickets)

Characteristics: this is one of the larger bush crickets. It is green with blackish markings on the body and wings. The female (see illustration) has a large ovipositor, and is thus easily ditinguished from the male, which produces a shrill chirp. Body length up to 35 mm.—**Occurrence:** common in meadows and healthland from July to October, living mostly on the ground. In Britain only recorded in a few southern counties.—**Diet:** the adults and nymphs feed on plants and small insects which they chew with their powerful jaws.—**Life history and habits:** the eggs are laid in the ground with the ovipositor and they spend the winter there. They hatch in spring and after a series of moults the nymphs are fully grown within a few months. The adult males then start to sing in order to attract the females. In addition to the green individuals there are also brown ones. The chirping song is produced by rubbing the wings against one another. The underside of the left wing has a toothed rib which is rubbed against the hind edge of the right wing. The sounds produced are high-pitched and rather prolonged. The song is characteristic for each species. Bush crickets are mainly nocturnal and most start to become active in the early evening. In parts of Europe it was once thought that a wart would disappear if it was bitten by one of these insects. This of course is quite untrue, and they are completely harmless. They usually escape by running or hopping but they can fly well. Their flight is not, however, comparable with that of the migratory locusts of the tropics which cause such damage to crops.

Praying Mantis *Mantis religiosa*

Order Mantodea (mantises)

Characteristics: grasshopper-like insects which do not, however, have hind legs adapted for jumping. On the other hand, their front legs are modified for catching prey. The tibia and femur are armed with spines on the inside and they can be closed together like a pocket knife, when catching prey. The prothorax is elongated and neck-like and the head is very mobile, giving the large eyes a wide visual field. When at rest the wings lie flat over the back, the larger hind wings being folded. The females are larger and stouter. Green and brown individuals occur. Body length up to 50 mm.—**Occurrence:** among vegetation, often at the edge of woodland, but only in the warmer, more southerly parts of Europe, and absent from Britain.—**Diet:** mantises are predatory, feeding on large prey, mainly insects which they cut in pieces with their powerful jaws. The nymphs also feed in this way.—**Life history and habits:** the eggs are laid in clumps and covered with a secretion which soon hardens to form the characteristic ootheca. The nymphs develop quite rapidly into the adult. The feeding drive is so strong that the smaller male is often eaten by the female during mating. Mantises live for a relatively long period of time and during her life a female may lay over a dozen times. The oothecae are found attached to rocks and twigs, where they remain for a long time after the eggs have hatched.

In the tropics there are many more mantid species, some of which have bizarre shapes, showing bright colours on the hind wings, and even mimicking flowers. They all have very similar habits and also makes use of their wings.

Common Earwig *Forficula auricularia*

Order Dermaptera (earwigs)

Characteristics: small, elongated brown insects with short front wings or elytra, which do not extend to the end of the abdomen. The caudal appendages are modified into powerful pincers which are more curved in the male than in the female. The membranous hind wings are much folded and hidden by the elytra and it is said that the pincers are used to fold the wings after flight. The antennae are long and thin. Body length up to 14 mm.—**Occurrence:** widely distributed and common on or near the ground, remaining hidden in crevices during the day. The adults spend the winter buried in the soil. Earwigs frequently enter houses.—**Diet:** mainly plants, often flowers but also small insects.—**Life history and habits:** the eggs are laid in the ground in autumn or spring and are guarded by the female. The young are also tended by the female, and they are fully grown by late summer. There is, of course, no truth in the legend that earwigs enter human ears and bite into the eardrum.

Oriental Cockroach (nymph)

Blatta orientalis (above left)

Forest Cockroach *Ectobius sylvestris* (above right)

Order Blattodea (cockroaches)

Characteristics: cockroaches are flattened insects with an oval outline, long antennae and walking legs all similar in form. They somewhat resemble beetles but are distinguished by their short caudal appendages. They are mostly an inconspicuous brownish colour. Body length of *E. sylvestris* 10 mm.—**Occurrence:** the Oriental Cockroach is only found in warm buildings as it comes originally from the tropics. The Forest Cockroach, on the other hand, lives on the ground in forests, particularly in mountainous areas. It is absent from Britain, where the related species *E. lapponicus* and *E. pallidus* occur in southern England.—**Diet:** plant matter including dead fragments. The species living in houses are pests feeding on household stores.—**Life history and habits:** cockroaches are nocturnal and spend the day hidden away. Using their powerful legs they can run very fast. The eggs are laid in a cocoon which is often carried around for a time by the female. The nymphs resemble the adults but have only short wing rudiments.

Green Shieldbug *Palomena prasina*

Order Hemiptera (true bugs)

Characteristics: small, flattened, roundish insects with a rostrum which is folded back against the ventral surface. They have moderately long antennae, normal walking legs and hard front wing covers which are, however, membranous at the tips. The prothorax is very large and triangular. The species is green but becomes dark bronze-red in autumn before it overwinters. In spring it becomes green again. Body length 14 mm.—**Occurrence:** common in bushes and trees, particularly on berried shrubs; widely distributed.—**Diet:** the adults feed on fruits to which they impart an unpleasant smell. The nymphs feed on plant sap.—**Life history and habits:** the eggs are laid on plants and the nymphs live there. They usually undergo 5 moults before reaching the adult stage. They and the adults contain symbiotic bacteria in their alimentary tract which produce additional vitamins. These insects have stink glands which produce a fluid giving the characteristic "buggy" smell.

Graphosoma italicum

Order Hemiptera (true bugs)

Characteristics: a bug with iridescent black and red coloration, the pattern being longitudinal on the back, but transverse on the belly. The legs and antennae are also conspicuously coloured. Body length 12 mm.—**Occurrence:** a warmth-loving species, living among vegetation. Not found in Britain.—**Diet:** plant sap and fruits.—**Life history and habits:** the eggs are laid in groups on plants, and after hatching the nymphs live gregariously for a time. Like the adults they are brightly coloured, and this probably serves as a warning because they are not edible, having an offensive taste. The secretion from the stink glands is slightly poisonous to other insects. Black and red warning patterns are found in many other unpalatable or poisonous insects. They appear to act as very effective warning signals to possible predators. Some harmless insects, lacking these secretions, also have this type of coloration; this is known as mimicry.

Firebug *Pyrrhocoris apterus* (opposite, above)

Order Hemiptera (true bugs)

Characteristics: a red and black bug with a longish-oval outline. The wing length is variable, as there are individuals with short and long wings. Body length 10 mm.—**Occurrence:** widely distributed but localized. Rare in Britain. Found living gregariously on limes and mallows.—**Diet:** adults and nymphs suck the sap of various plants.—**Life history and habits:** the adults overwinter and appear again in the following spring when the eggs are laid in large groups. Bug eggs vary considerably in shape (opposite, below right). The nymphs grow during the course of the summer and like the adults they live in large aggregations. The illustration shows a pair mating.

Rhinocoris iracundus (opposite, below left)

Order Hemiptera (true bugs)

Characteristics: a red and black bug with the head elongated at the front, and with slightly concave sides. Body length 12 mm.—**Occurrence:** a widely distributed species, mostly living singly among vegetation. Not found in Britain.—**Diet:** both adults and nymphs are predators, which feed on other insects, using the piercing rostrum to suck their tissues.—**Life history and habits:** the nymphs develop in the same way as those of other bugs. The rostrum injects saliva which paralyses and kills the prey. The rostrum is also used in defence.

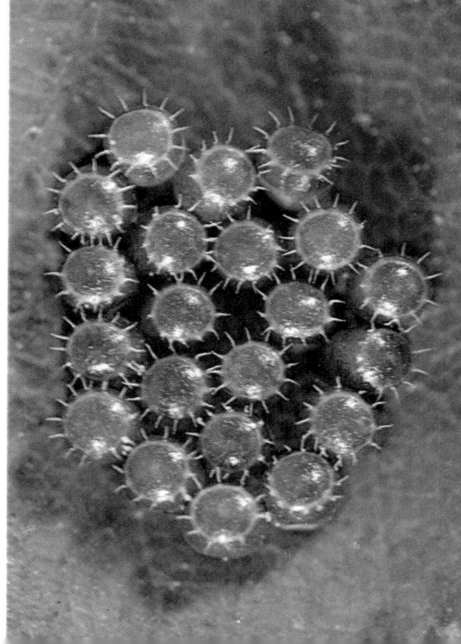

Common Pondskater *Gerris lacustris*

Order Hemiptera (true bugs)

Characteristics: medium-sized, dark bugs with very long legs with support the insect on the water surface. The front legs are much shorter than the second pair of legs. Body length 12 mm.—**Occurrence:** widely distributed on quiet waters, often living in large groups.—**Diet:** these are predators which seize prey on the surface of the water, mainly taking small insects.—**Life history and habits:** the nymphs develop like those of other bugs. Pondskaters row themselves along on the water surface with the middle legs, steer with the hind legs and seize prey with the front legs. The adults can fly well and in this way they colonise new areas of suitable water.

Common Water-boatman *Notonecta glauca*

Order Hemiptera (true bugs)

Characteristics: bugs with a broader body than pondskaters, and with long, powerful hind legs fringed with swimming hairs. These bugs are aquatic and they always swim back upwards. Like most bugs they have wings and fly well. Body length 15 mm.—**Occurrence:** abundant throughout most of the year in standing waters.—**Diet:** these are predators which take prey in the water and at the surface, seizing it and piercing it with the rostrum.—**Life history and habits:** the nymphs develop like those of other bugs. In the water water-boatmen breathe from an air bubble held on the underside of the abdomen. They have to come to the surface at frequent intervals in order to renew this air supply. As they are predators they should not be kept in an aquarium where they will even attack small fishes. When handled they will bite in self-defence and this is very painful.

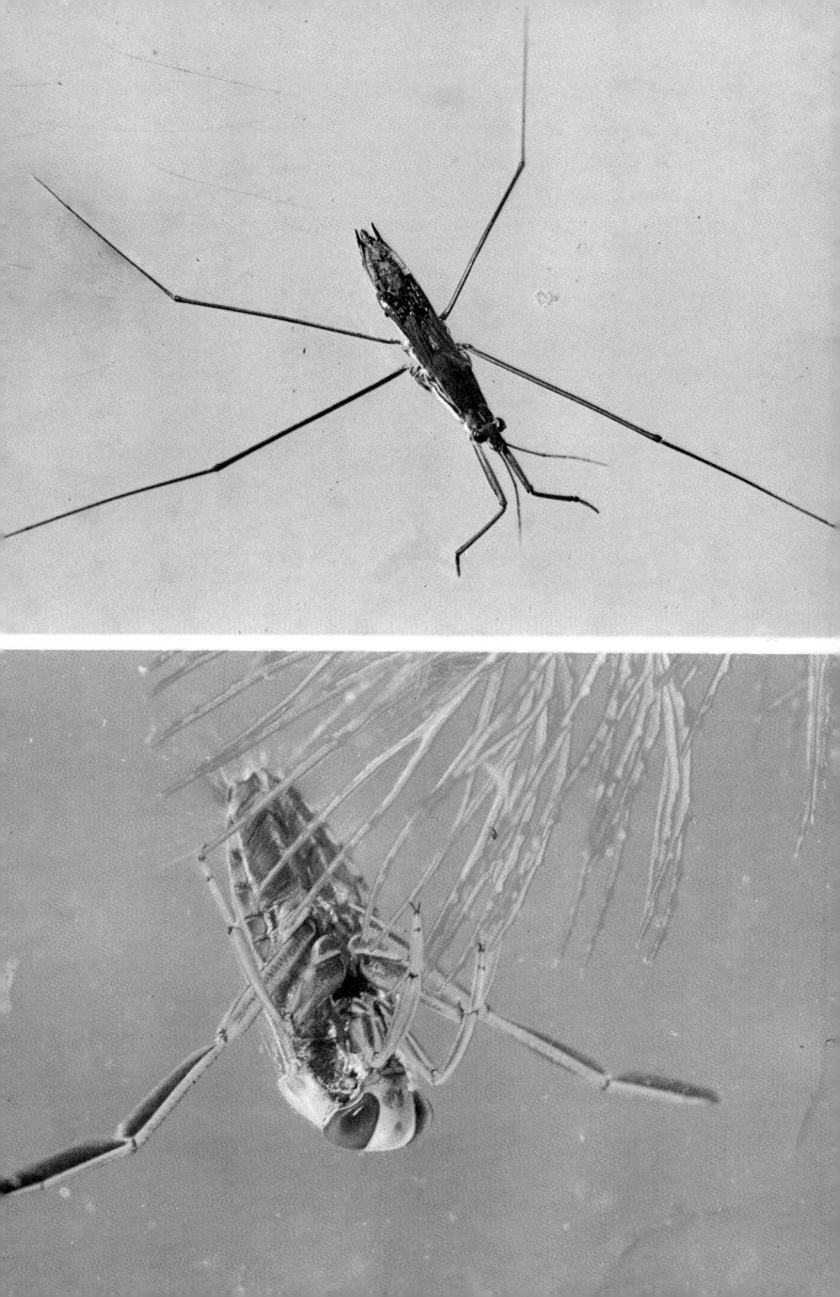

Cuckoo-spit Insect *Philaenus spumarius* (opposite, above)

Order Hemiptera (true bugs)

Characteristics: small, brownish bugs which are very inconspicuous. The pale green nymph is shown here with its mass of froth. These are commonly seen on plant stems.—**Occurrence:** widely distributed in fields and meadows, where the nymphs and their froth are very conspicuous, but the adults are difficult to find.—**Diet:** plant juices.—**Life history and habits:** the nymphs blow air from the end of the abdomen into a liquid excreted at the anus, and this produces the protective froth, known colloquially as cuckoo spit.

Froghopper *Cercopis vulnerata* (opposite, below left)

Order Hemiptera (true bugs)

Characteristics: small, oval insects with the wings held rooflike over the body. They have a pattern of black and red. The powerful legs enable these insects to leap for relatively long distances, and they also fly well. Body length 6 mm.—**Occurrence:** widely distributed and common among low vegetation in open country.—**Diet:** both adults and nymphs feed on plant juices which they suck from the tissues with the beak or rostrum.—**Life history and habits:** the nymphs live underground on roots and climb up into the light when they emerge as adults.

Cicadetta montana (opposite, below right)

Order Hemiptera (true bugs)

Characteristics: a large brownish cicada, about 20 mm long, with large transparent wings. Cicadas have a very broad head with widely separated eyes and very short antennae. The males produce sounds to attract the females.—**Occurrence:** only in the warmer parts of Europe and never abundant. In Britain only found in the New Forest.—**Diet:** plant juices.—**Life history and habits:** the nymphs live in the ground, usually for several years. The sounds are produced by two drum-like membranes, the tymbals, one on each side of the abdomen. The illustration shows a recently emerged specimen with the wings not yet fully developed.

Greenfly *Aphis* sp.

Order Hemiptera (true bugs)

Characteristics: very small insects with membranous wings and a suctorial rostrum. The head is pointed and the abdomen broad and rounded. In addition to the winged, flying individuals there are also flightless forms which occur in large numbers.—**Occurrence:** greenflies and other aphids live on plants. Certain species are adapted for living on a definite species of plant.—**Diet:** all stages suck plant sap.—**Life history and habits:** the eggs overwinter and hatch in spring to produce flightless females each of which, without mating, produces another flightless female about once a day; this is known as parthenogenesis. Winged individuals appear from time to time and these disseminate the species. Males appear in autumn and mate with females which then lay eggs. In some species this alternation of generations is dependent upon the food plant, the sexual and parthenogenetic individuals living on different plants. On account of the rapid rate of multiplication greenfly and related aphids are often serious pests.

Alderfly *Sialis* sp.

Order Neuroptera, sub-order Megaloptera (alderflies and snakeflies)

Characteristics: large, darkish insects with the wings longer than the body and with reticulate veins. Antennae long, mouthparts biting. Distinguished from the similar stoneflies by the absence of caudal appendages, and from the caddisflies by having the wings almost equal in size.—**Occurrence:** in the vicinity of water, often among vegetation. Widely distributed.—**Diet:** the aquatic larvae are predators that feed on insects; adults scarcely feed at all.—**Life history and habits:** metamorphosis is complete. The larvae breathe by 7 pairs of feathery tracheal gills on the abdomen. After a year or more the larvae creep on to the land and pupate in the soil. On emergence from the pupal stage the adults mate and the females lay eggs in large batches near water and the newly hatched larvae drop or crawl into the water.

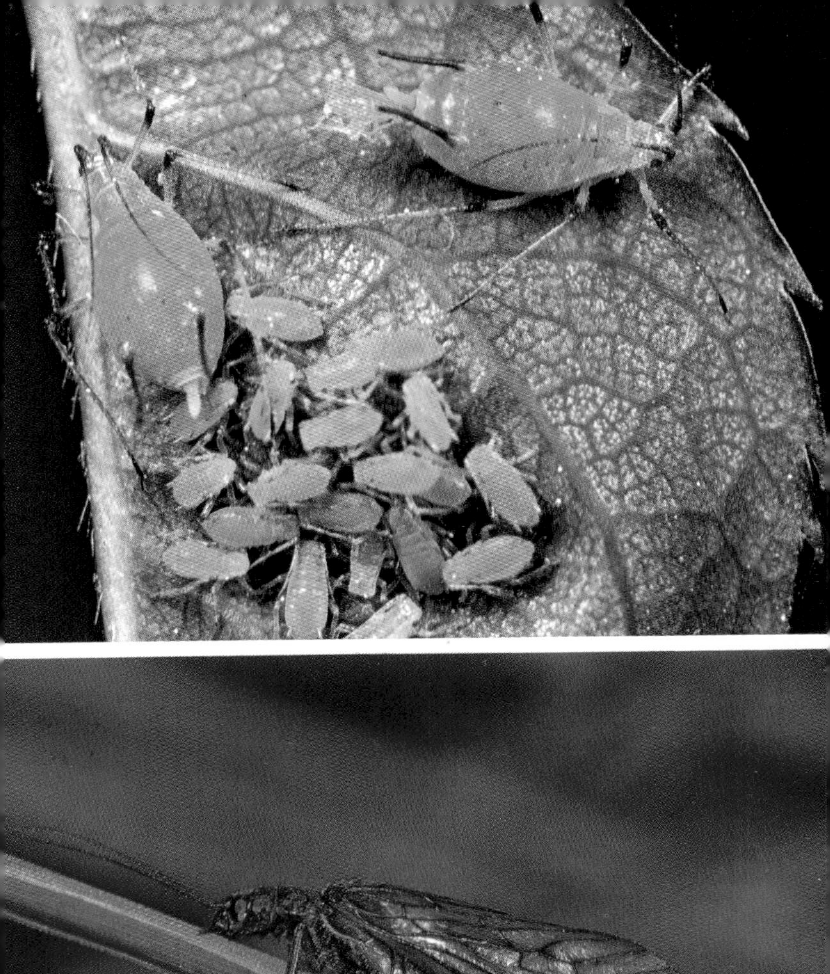

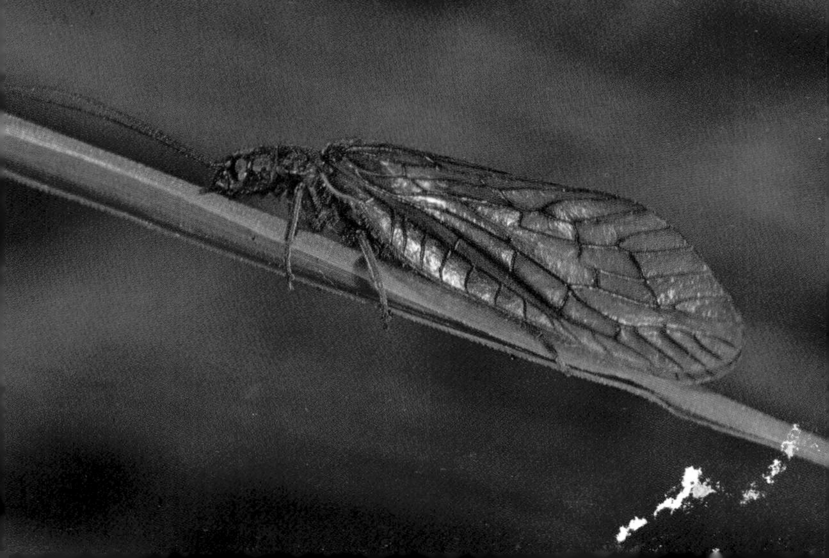

Snakefly *Raphidia* sp.

Order Neuroptera, sub-order Megaloptera (alderflies and snakeflies)

Characteristics: small terrestrial insects with an elongated body and a very long, neck-like prothorax. Head with biting mouthparts and long antennae. The membranous wings are all similar in shape, with a network of veins. Females have a long ovipositor. Body length 10 mm.—**Occurrence:** locally in woodland in summer, but usually not common.—**Diet:** all stages are predatory, the larvae living in the crevices of bark, the adults on tree trunks.—**Life history and habits:** larval development takes about two years, and there is then a pupal stage. After mating the females lay their eggs in bark crevices. The larvae can be regarded as beneficial to man as they hunt wood-boring insects.

Lacewing *Chrysopa* sp.

Order Neuroptera, sub-order Planipennia (true lacewings)

Characteristics: small insects with a slender, iridescent body and long antennae. The wings are large, membranous and similar in shape, with the longitudinal veins connected by numerous transverse veins, forming a lace-like network; the veins are coloured, usually green. The eyes are iridescent. Body length 14 mm.—**Occurrence:** widespread and common among vegetation in summer and autumn, particularly if aphids are present; sometimes seen in houses during the winter.—**Diet:** the larvae feed on aphids which they pierce and suck.—**Life history and habits:** the eggs are laid at the end of long, thin mucus threads, which form stalks. The larvae live on plants, feeding on aphids, and are therefore beneficial to man. The adults are active at night and are attracted to the light. When theatened they produce an evil-smelling substance and are therefore avoided by their enemies. They overwinter and their colour then changes from green to pinkish, but back again to green in the following spring.

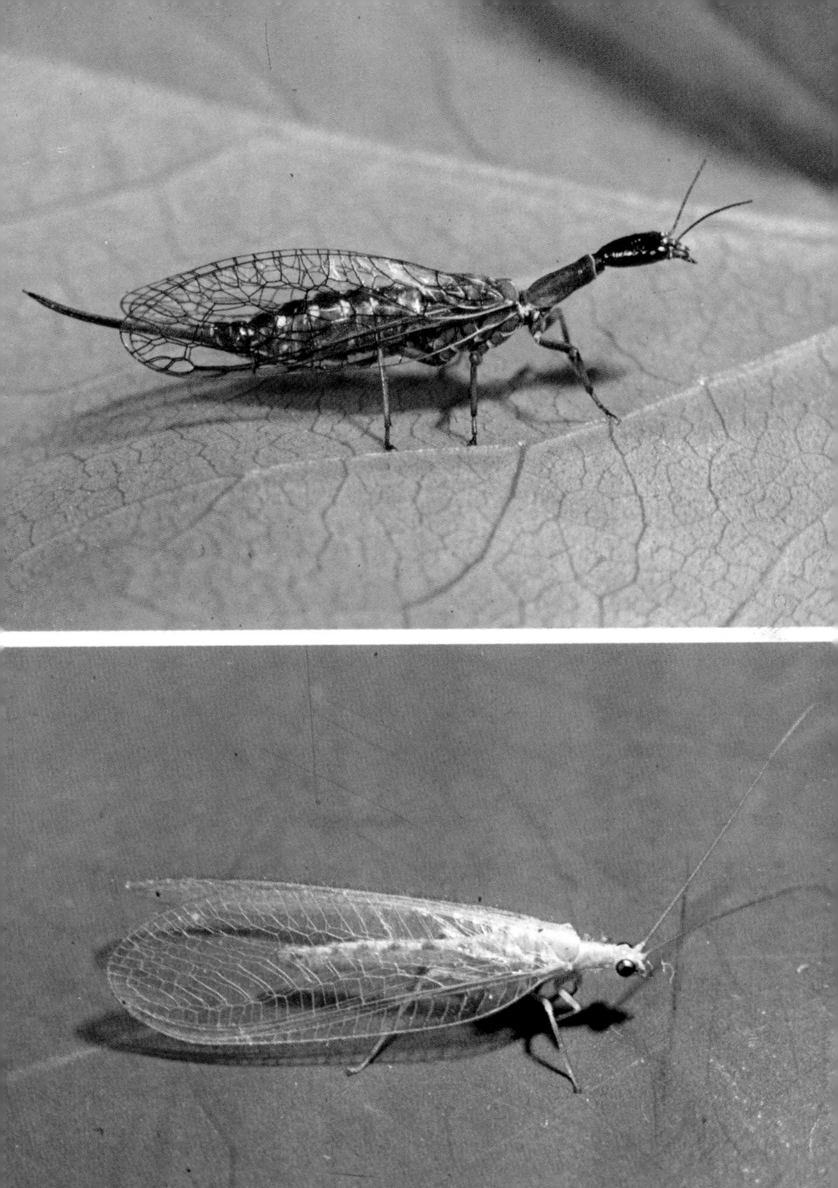

Ant-lion *Myrmeleon formicarius*

Order Neuroptera, sub-order Planipennia (true lacewings)

Characteristics: the adults (opposite, above) have a long, thin body and large membranous wings, thus somewhat resembling a dragonfly. They differ, however, in having short, club-like antennae and dull coloration, as well as some hair on certain parts of the body. The adults are not often seen. On the other hand, the larvae (opposite, below) are much better known, and indeed the name ant-lion really refers to them. They have a roundish grey-brown body with strong jaws. Body length of the adult 70 mm, of the larva 15 mm.—**Occurrence:** widespread but local in warm, sandy areas. Not found in Britain.—**Diet:** the larvae mainly catch ants, and also other insects. Little is evidently known about the diet of the adults; it is possible that they do not feed.—**Life history and habits:** metamorphosis is complete, with a pupal stage. The larva lies at the bottom of a funnel-shaped pit in sandy soil, with only the jaws above the surface, lying in wait for prey. When a small insect falls into the funnel it finds difficulty in climbing out. The more it tries to escape the more it is bombarded by sand grains so that it falls back again into the pit. There it is seized by the larva's long hollow jaws which pierce its body and suck its tissues. The empty husks are thrown out of the pit. If the funnel is in any way damaged it is immediately repaired. The adults are active by day, particularly in sunlight, but only fly for short distances with an undulating flight path. In the tropics there are several very large species, which mainly fly by night and these are attracted by light.

Ascalaphus libelluloides

Order Neuroptera, sub-order Planipennia (true lace-wings)

Characteristics: butterfly-like lace-wings with a slender, hairy body, long black and white wings and long clubbed antennae. Body length 60 mm.—**Occurrence:** in open woodland, mainly in the southern part of Europe, but not very common. Not found in Britain.—**Diet:** the larvae and adults are predatory.—**Life history and habits:** the larvae live on the ground where they hunt, seizing their prey with the suctorial jaws. They resemble ant-lion larvae, but do not build a pit trap. The adults fly about very rapidly and catch flies and butterflies on the wing.

Caddisfly *Anabolia nervosa*

Order Trichoptera (caddisflies)

Characteristics: medium-sized, inconspicuously coloured insects (opposite, below right) with membranous hairy wings with few veins. When at rest the wings are folded together over the body, rather like a roof. The antennae are very long and the mouthparts are biting, but usually not well developed. The adults are distinguished from similar insects in other orders by the absence of caudal appendages, by the hairy wings and by the small number of transverse veins. The larvae are aquatic, each living in a portable case (opposite, below left) made of sand grains or plant material. Each species has a favourite method of constructing its case. Body length 30 mm.—**Occurrence:** always in the vicinity of quiet waters, among vegetation by day, flying in the evening and at night.—**Diet:** the larvae feed on plant matter and invertebrates. Some species build fixed silken nets which catch food in fast-flowing waters. The adults scarcely feed at all.—**Life history and habits:** the larval cases are open at both ends, thus allowing water to circulate, so that oxygen reaches the feathery gills on the sides of the body. They pupate in the case, which is fixed to some firm object. Those that do not live in a case make a pupal chamber in the sand. The pupa continues to obtain oxygen through the gills. Just before emergence it bites its way out of the pupal case and moves to the surface. There it emerges as an adult and flies off.

Scarce Swallowtail *Iphiclides podalirius*

Order Lepidoptera (butterflies and moths)

Characteristics: a broad-winged butterfly with a relatively small body. The wings are pale yellow with transverse black bars. Each hind wing has a "tail". Antennae long and clubbed. Mouthparts modified to form a suctorial proboscis which can be coiled under the head. Wing span c. 80 mm.—**Occurrence:** local and uncommon in central Europe, with one generation in May-June. In southern Europe there are two broods (spring and summer). Not found in Britain, except as a rare vagrant.—**Diet:** the caterpillars feed on the leaves of sloe, apricot and related trees, the adults suck flower nectar.—**Life history and habits:** the adults fly around in hilly areas where the food plants of the caterpillars grow, and on which the eggs are laid. The caterpillars grow for a few months, then pupate and overwinter in this stage. On emergence the wings are still quite small and soft (opposite, above left). They expand to their full size as blood is pumped through the veins, and then harden in the air. This process scarcely takes on hour and the butterfly can then fly.

Common Swallowtail *Papilio machaon*

Order Lepidoptera (butterflies and moths)

Characteristics: somewhat similar to the preceding species but the wings are darker yellow with a black reticulate pattern. Hind wings with a "tail" and a red and blue eye-spot at the inner angle. Wing span 60-80 mm.—**Occurrence:** widely distributed in Europe, but in Britain only in Norfolk. Two broods (spring and summer).—**Diet:** the adults visit flowers to suck nectar. The caterpillars feed on umbelliferous plants, such as dill, caraway and fennel.—**Life history and habits:** the fully grown caterpillars pupate in autumn and overwinter in this stage, which is brown. The adults of the spring brood produce caterpillars which form summer pupae which are usually green. After 2-3 weeks these emerge as the summer brood. The males often gather on hilltops and fly around waiting for the females who visit such places to find a mate.

Orange Tip *Anthocaris cardamines*

Order Lepidoptera (butterfiles and moths)

Characteristics: a medium-sized butterfly with white wings, a black spot on each front wing and black wing tips. In the males the outer half of each front wing is orange, so the sexes are easy to distinguish. Wing span 30-35 mm.—**Occurrence:** a fairly common butterfly in open country.—**Diet:** the adults visit flowers for nectar, the caterpillars feed on cress, hedge mustard and other cruciferous plants.—**Life history and habits:** the pupae overwinter attached to plants and are well camouflaged by their coloration. The adults emerge and have a life of several weeks, during which the eggs are laid on the food plants. The caterpillars feed for about two months and then pupate.

Brimstone *Gonepteryx rhamni*

Order Lepidoptera (butterflies and moths)

Characteristics: a large butterfly, all the wings with sharp angular points. The males are bright lemon-yellow, the females yellowish-white, both with an orange spot on the centre of each wing. Wing span c. 60 mm.—**Occurrence:** widespread in open country close to woodland and not uncommon. Adults are seen on the wing (see below) during three periods of the year: March-April, July and September.—**Diet:** the butterflies suck nectar, the caterpillars feed on buckthorn.—**Life history and habits:** the adults live a long time, sometimes up to 9 months. This is associated with their characteristic life cycle. The overwintering adults are on the wing during spring when the females lay eggs. These hatch into caterpillars which pupate to give the next brood about three months later. In fact these new adults are seen flying during July. In some areas there is another brood flying in August and September. The related Cleopatra (*G. cleopatra*) occurs in the Mediterranean area.

Large White *Pieris brassicae*

Order Lepidoptera (butterflies and moths)

Characteristics: a large white butterfly with black wing tips and in the females two black spots on each front wing. Wing span 60-70 mm.—**Occurrence:** the adults appear in spring and there are usually two broods between then and the autumn. They are very common in open country, meadows and gardens.—**Diet:** the adults suck nectar, the caterpillars feed on cruciferous plants and when present in large numbers may cause serious damage to crops.—**Life history and habits:** the adults seen in spring have emerged from overwintering pupae and they look for wild crucifers. The summer adults are considerably commoner and they now lay their eggs on cultivated crucifers, wandering round in large numbers from field to field. The caterpillars are often parasitized by braconids (described and illustrated on p. 110), and this much reduces their numbers.

Painted Lady *Vanessa cardui*

Order Lepidoptera (butterflies and moths)

Characteristics: a large brown butterfly with black and white markings which is well camouflaged when resting on the ground. Wing span 50-60 mm.—**Occurrence:** the adults appear in variable numbers from May to October, Mainly in open country.—**Diet:** the butterflies visit flowers for nectar, the larvae feed on thistles and stinging nettles.—**Life history and habits:** this species is really an inhabitant of tropical and subtropical steppe country and so cannot survive over winter in northern and central Europe. At certain seasons the vegetation in its home range becomes scarce owing to drought and it then migrates. With favourable wind conditions it may reach the Alps and the Mediterranean and occasionally as far as Iceland. It is, therefore, not surprising that this species has been recorded in all parts of the world, except South America. Its power of flight is astonishing.

Peacock Butterfly *Inachis io*

Order Lepidoptera (butterflies and moths)

Characteristics: a large, red-brown butterfly easily distinguished by the large eye-spots on each wing. The front wings are distinctly notched on the outer edge. The underside of the wings is greyish-black and so provides good camouflage when the insect is at rest with the wings folded over the back. Wing span 50-60 mm.—**Occurrence:** widespread and very common in open country and along the edges of woodland. There are two broods in the year.—**Diet:** the adults take nectar, the caterpillars feed on stinging nettles.—**Life history and habits:** adults overwinter and come out with the first sunshine of the spring. They lay eggs on nettles on which the resulting caterpillars live gregariously. Although these live in a loose web and carry numerous long spines they are frequently decimated by parasitic Hymenoptera. The offspring of the summer brood appear in August and overwinter as adults.

Red Admiral *Vanessa atalanta*

Order Lepidoptera (butterflies and moths)

Characteristics: a large butterfly with dark wings marked with a brilliant red and white pattern. The outer edges of the front wings are angular. The underside has a marbled pattern which provides excellent camouflage. Wing span 50-60 mm.—**Occurrence:** in open country and often in gardens, from May to October. This is a migratory species whose numbers fluctuate.—**Diet:** the adults suck nectar, and in autumn they are also attracted by rotting windfalls. The caterpillars mostly feed on stinging nettles.—**Life history and habits:** the adults seldom overwinter in central Europe and Britain, but almost every year varying numbers migrate northwards from the Mediterranean area. In autumn some of the adults may try to return south, but at this season they can scarcely survive the flight over the Alps. Those that remain to the north will continue to fly until late autumn but will die with the arrival of the first severe frosts.

Small Tortoiseshell *Aglais urticae*

Order Lepidoptera (butterflies and moths)

Characteristics: a medium-sized red-brown butterfly with angular wings. The red-brown ground colour is marked with a pattern of black, yellow and white, with a row of blue spots around the edges of the wings. The underside is blackish-brown and when the insect is at rest this provides good camouflage. Wing span 35-50 mm. The caterpillar is brownish-black with yellow longitudinal stripes and long spines. Depending upon the background the pupae are pale brown to blackish with iridescent gold spots. They hang head downwards, held by the tip of the abdomen.—**Occurrence:** widely distributed and common in open country and woodland clearings. Overwintering adults are frequently seen in houses, particularly in roof-spaces. These insects appear on the first warm days of spring, but if cold weather returns they will go back into hibernation.—**Diet:** the adults take nectar and also tree sap. The caterpillars (opposite, above right) feed on stinging nettles.—**Life history and habits:** the caterpillars live more or less gregariously in a loose web and do not disperse until the last stage. Although they are protected by their food plant and by the numerous spines on the body they are still much parasitized by parasitic Hymenoptera. When fully grown the caterpillars look for a suitable place in which to pupate. The pupa hangs head downwards (opposite, above left), often from walls, posts and tree trunks. Although the caterpillars are very abundant the pupae are very seldom seen; their colours match the background and this protects them from enemies. After three weeks the adult or imago emerges and lives for several weeks. The adults, and particularly the females, require some nourishment in the form of nectar for the eggs to mature. These are laid in clumps of 50-100 on nettle. This life cycle is characteristic of all butterflies in which the adults feed. They only differ in the number of broods and in the developmental stage which overwinters.

Map Butterfly *Araschnia levana*

Order Lepidoptera (butterflies and moths)

Characteristics: a small butterfly with very attractive markings. In the first or spring brood (opposite, above) the ground coloration is red-brown, in the second or summer brood (opposite, below) it is black. The two forms were at one time described as separate species. Wing span 30-40 mm.—**Occurrence:** widely distributed but local along the edges of woodland and in clearings. Not found in Britain. The adults fly in May-June and August-September.—**Diet:** the adults suck nectar, the larvae feed gregariously on nettles.—**Life history and habits:** the existence of the two colour variants is associated with the type of development. The offspring of the summer generation pupate in autumn and the pupae overwinter. This overwintering is brought on by the shorter day lengths in autumn. If these pupae are exposed experimentally to more than 14 hours daylight in the twenty-four, their development will continue and they will emerge as black "summer" individuals. The overwintering pupae required a period of frost followed by warmth before metamorphosing and emerging as red-brown "spring" individuals. Their offspring, developing in the long days of summer, have a brief pupal phase and emerge as black "summer" individuals. This difference, known as seasonal dimorphism, is found quite commonly in tropical butterflies, where it is mainly associated with the dry and rainy season. In the Map Butterfly the seasonal difference in colour is primarily dependent upon the day length, and also on the temperature.

In this species the method of egg-laying is of interest. The eggs, which look like small barrels, are attached to the food plant one above the other, forming little towers. The caterpillars hatching from the outermost eggs have to climb across the other eggs in order to reach the leaf. The caterpillars are very similar to those of the Small Tortoiseshell (p. 69, above right) and they pupate in the same way (p. 69, above left), hanging head downwards.

Silver-washed Fritillary *Argynnis paphia*

Order Lepidoptera (butterflies and moths)

Characteristics: a large species in which the male is bright red-brown, the female olive-brown. The pattern consists of black spots and streaks. The male has scent scales which are absent in the female. These appear as black stripes which run along the veins on the centre of each front wing. The underside of the hind wings is greenish with silvery markings which look rather like mother-of-pearl. In other related species there are iridescent silvery spots. Wing span 60-80 mm.—**Occurrence:** widely distributed in the vicinity of woodland and especially in clearings, where they visit thistles in particular during summer. There is one brood a year.—**Diet:** the adults suck flower nectar. The caterpillars feed mostly on species of *Viola* and remain hidden by day.—**Life history and habits:** the eggs are laid in July in the vicinity of violet plants and hatch in August. The caterpillars feed for a brief period and then go into hibernation, waking up in the following spring. The pupae hang upside down attached to vegetation. The males perform a complicated courtship in which the scent scales play a part. These have gland cells at their base which produce a scent which spreads over the scent scales. At a certain point in the courtship when the female's antennae come to lie betwen the male's wings she picks up the scent with her antennae and is then ready to mate. Scents, in fact, play an important role in the life of butterflies and moths. They lead them to their food sources and guide the female to the right place for egg-laying. This is an extremely important function because most caterpillars only eat certain plants and, if these are not available, they will die rather than eat anything else. The caterpillars also find the correct food plant by scent, and can, in fact, be induced to feed on paper which has been soaked in an extract of the original food plant.

Marbled White *Melanargia galathea*

Order Lepidoptera (butterflies and moths)

Characteristics: a medium-sized butterfly with a very characteristic pattern of angular black markings on a white background. The wings are rounded. The underside is lightly patterned with circular spots on the hind wings. Wing span 40-50 mm.—**Occurrence:** widespread on grassland, meadows and heathland, with one brood a year.—**Diet:** the adults suck the nectar of flowers such as thistles. The caterpillars feed on grasses, remaining hidden by day.—**Life history and habits:** the eggs are laid low down and the caterpillars overwinter when still very small. They pupate in the following year, the pupae hanging head downwards. The adults fly relatively slowly, coming down to suck nectar quite frequently. They are sometimes seen in the cool of the morning covered with dew and sitting motionless on flowers.

Arran Brown *Erebia ligea*

Order Lepidoptera (butterflies and moths)

Characteristics: a medium-sized, dark brown butterfly with red-brown bands enclosing three or four small black eye-spots. The wing edges have a black and white chequered pattern. The underside of the hind wings has white spots. Wing span 40-50 mm.—**Occurrence:** one brood a year in open woodland. Not found in Britain, for in spite of its popular name there is no good evidence that it has ever occurred in Arran.—**Diet:** the adults visit flowers and the caterpillars feed on grasses.—**Life history and habits:** the caterpillars overwinter and pupate in the ground.

There are several species of *Erebia* which are all very similar. These occur up to great heights in mountainous areas. In such places local races have evolved which differ from one another in the size of the brown bands and in the development of the eye-spots. They are often seen in large numbers in alpine pastures.

Green Hairstreak *Callophrys rubi*

Order Lepidoptera (butterflies and moths)

Characteristics: a small butterfly which is uniformly dark brown on the upperside. The underside is bright green, the hind wings marked with a few small white spots. The club-shaped antennae are conspicuously ringed in black and white. The hind wings have short "tails". Wing span 25 mm.—**Occurrence:** flying rapidly in woodland clearings and along the edges of forests. One brood a year. —**Diet:** the adults visit flowers to suck nectar. The caterpillars feed on various plants, such as raspberry, bilberry and broom.—**Life history and habits:** the caterpillars, shaped somewhat like a wood-louse with an arched, hairy upper surface, pupate in the autumn. The pupae are attached to various objects at the hind end and also by means of a girdle round the middle of the body. There are other hairstreaks in which the caterpillars are dark with a whitish ring round the middle. This gives them the appearance of a bird dropping and provides excellent camouflage when they are on the surface of a leaf. When the butterfly is at rest the green underside of the wings also serves as camouflage and it is extremely difficult to see. As in the Silver-washed Fritillary there are scent scales on the upperside of the front wings and these play an important part in courtship.

In the tropics there are several hairstreaks which are often brilliantly coloured. This is due to iridescent scales which owe their colour not to pigments but to reflection from numerous thin lamellae in the scales, a phenomenon known as interference. Iridescent scales are mainly found in the male and may serve as signals to the opposite sex. These scales are mostly coloured blue, red or violet. The iridescence of the Green Hairstreak also arises in this way although it is not so pronounced as it is in the following two species.

Common Blue *Polyommatus icarus*

Order Lepidoptera (butterflies and moths)

Characteristics: a small butterfly with an iridescent blue upperside and narrow black wing edges fringed with white. The underside is grey-brown with black spots encircled by white and with a row of orange spots along the edge. The females have a brown upperside with a few blue and orange spots. Wing span 25 mm.—**Occurrence:** widespread in meadows and on hillsides. There are usually two broods between May and September.—**Diet:** the adults visit flowers to take nectar, the caterpillars feed on leguminous plants, such as clovers.—**Life history and habits:** the half-grown caterpillars overwinter and pupation takes place in the following spring. There are several other similar blues, which are often very difficult to distinguish. They all have more or less conspicuous coloration on the upperside and small spots on the underside. The caterpillars of some species live in association with ants.

Scarce Copper *Heodes virgaureae*

Order Lepidoptera (butterflies and moths)

Characteristics: a small, iridescent golden butterfly with a narrow black edging to the wings. The underside is grey-brown with small dark dots. In the females the upperside is darkened by brownish markings. Wing span 30 mm.—**Occurrence:** at one time widespread in damp meadows, but now very scarce owing to the spread of cultivation. The species does not occur in Britain. One brood in the year.—**Diet:** the adults suck nectar, while the caterpillars feed on various species of dock.—**Life history and habits:** the eggs are laid on the food plant where they overwinter. They hatch in April and the caterpillars start to feed, eventually pupating in June.
The brilliant iridescent coloration is due to the phenomenon of interference described above. If a drop of alcohol is dropped on the wing the iridescence disappears immediately, but re-appears when the liquid has evaporated. These colours are doubtless important as signals to the opposite sex.

Elephant Hawk-moth *Deilephila elpenor*

Order Lepidoptera (butterflies and moths)

Characteristics: a large hawk-moth in which the front wings are pink and olive-green, the hind wings pink and dark brown. The abdomen is pink with areas of olive and a few dark markings. Body length 30 mm, wing span 60-70 mm.—**Occurrence:** common in gardens and parks and in open country, where it flies during the evening and at night in May-June.—**Diet:** the adults suck nectar, the caterpillars feed on bedstraws, willowherb and fuchsia.—**Life history and habits:** hawk-moths have very large caterpillars which are naked and usually carry a curved horn on the abdomen. They appear somewhat frightening but are completely harmless. In addition the caterpillars of the present species and of related ones have eye-spots in front which give them a snake-like appearance. The pupae overwinter in the ground.

Privet Hawk-moth *Sphinx ligustri*

Order Lepidoptera (butterflies and moths)

Characteristics: a large hawk-moth with narrow wings, a stout body and thick antennae. The front wings have a brownish watered pattern, while the smaller hind wings are pink with dark bands. The thorax is mainly dark brown, the abdomen pink with dark transverse bars. Body length 40 mm, wing span about 100 mm.—**Occurrence:** widely distributed and quite common in gardens, parks and along the edges of woodland during June and July, flying in the evening and at night.—**Diet:** the adults regularly suck the nectar from flowers of lilac or honeysuckle. The caterpillars feed on privet, lilac and ash.—**Life history and habits:** when sucking nectar with their very long proboscis the adults hover in the air, as though motionless. This behaviour pattern is particularly characteristic of the hawk-moths. They fly very fast, perhaps faster than any other lepidopterans.

Spurge Hawk-moth *Celerio euphorbiae*

Order Lepidoptera (butterflies and moths)

Characteristics: a large moth with pale brown front wings marked with dark brown bars and flecks. The hind wings are reddish with dark bands. The abdomen is yellowish with an olive-brown and black pattern. Body length 30 mm, wing span 60 mm.—**Occurrence:** a migratory moth, mainly living in the Mediterranean area. This species occurs in varying numbers in the drier and warmer areas further north, including Britain. There are two broods a year (spring and summer) and the adults fly in the evening and at night.—**Diet:** the adults suck nectar, the larvae feed on spurges. —**Life history and habits:** the attractive caterpillars are very active during the day on their food plant. Their bright coloration probably serves as a warning that they are unpalatable. They store poisonous substances from the plants in their body, and attackers such as birds learn to avoid them. Such protective mechanisms are very common in butterflies and moths.

Poplar Hawk-moth *Laothoe populi*

Order Lepidoptera (butterflies and moths)

Characteristics: a large grey-brown moth with some green areas. The wing edges are wavy. The hind wings are copper-coloured at the base. In this species the proboscis is reduced. Body length 25 mm, wing span 60-90 mm.—**Occurrence:** common along the edges of woodland and in parks and gardens, flying at night. There are two broods between May and September.—**Diet:** the adults scarcely feed at all, as the proboscis is reduced. The larvae feed on the leaves of poplar and willows.—**Life history and habits:** the resting position is quite characteristic. The hind wings are drawn far forward and they overlap the front wings which are pulled backwards. Together with the inconspicuous coloration this provides excellent camouflage, when the moth is resting by day among vegetation. The pale green caterpillar has lateral yellow stripes and is thus almost invisible when on willow leaves.

Eyed Hawk-moth *Smerinthus ocellata*

Order Lepidoptera (butterflies and moths)

Characteristics: a large, mainly red moth with a brownish pattern on the front wings and an eye-spot encircled in black on each hind wing. The body is brownish, the dorsal side of the thorax having a dark brown band. Body length 25 mm, wing span 60-90 mm.—**Occurrence:** flying at night, and common in parks and along the edges of woodland from May to July.—**Diet:** the adults have reduced mouthparts and scarcely feed at all. The caterpillars feed on poplar and willow.—**Life history and habits:** during the day these moths remain well camouflaged among vegetation. If threatened, however, the hind wings are exposed and the large eye-spots become visible. The attacker is frightened because it seems to see a much larger adversary looking at it with big eyes.

Broad-bordered Bee Hawk-moth

Hemaris fuciformis

Order Lepidoptera (butterflies and moths)

Characteristics: a smaller moth with transparent wings that have brown edges. The body is olive-green, brown, yellow and black and resembles a bumblebee. Body length 15 mm, wing span 40 mm.—**Occurrence:** widely distributed and common in meadows and along the edges of woodland. It flies by day when the sun is shining and visits flowers.—**Diet:** the adult moths suck nectar, the caterpillars feed on honeysuckle leaves.—**Life history and habits:** the species overwinters in the pupal stage. The adult or imago emerges in spring. When visiting flowers it can hover in the air and this and the almost silent flight help to distinguish it from a bumblebee. An attacker might well mistake it for such a stinging insect, and this is another case of mimicry. The tropics are particularly rich in butterflies and moths showing mimicry.

Hummingbird Hawk-moth *Macroglossum stellatarum*

Order Lepidoptera (butterflies and moths)

Characteristics: a small moth with grey-brown front wings crossed by thin darker lines and yellow-brown hind wings. The body is greenish-brown. Body length 15 mm, wing span 40 mm.—**Occurrence:** a migratory insect which only visits northern Europe and Britain at irregular intervals. It flies during the day in open country.—**Diet:** the adults visit flowers, particularly jasmine, for their nectar. The caterpillars feed on bedstraws.—**Life history and habits:** those that arrive by migration produce a brood but these offspring do not survive the winter in Britain. In southern Europe this species produces two broods in the year.

Scotch or Mountain Burnet

Zygaena exulans

Order Lepidoptera (butterflies and moths)

Characteristics: small moths with long clubbed antennae. The front wings are greenish-black with red markings, the hind wings are red with a black border. Wing span 30 mm.—**Occurrence:** the species illustrated only occurs in mountainous areas, but several related species occur very commonly in meadows during the summer.—**Diet:** the adults suck nectar, especially of scabious and thistles. The caterpillars feed on leguminous plants, such as clovers and trefoils.—**Life history and habits:** there is usually only one brood in the year. The young caterpillars overwinter. When fully grown they construct papery yellow cocoons which are attached to grass stems. Before the emergence of the adult in July the pupa protrudes slightly from the cocoon.
Both the adults and the caterpillars are unpalatable and they advertise this by their warning colours, which are red and black in the adult, yellow and black in the caterpillar.

Emperor Moth *Saturnia pavonia*

Order Lepidoptera (butterflies and moths)

Characteristics: a large moth with an eye-spot on each wing. The front wings are grey-brown and the hind wings yellowish in the male (opposite, above), but in the female all four wings are grey (opposite, below). The antennae are feathery in the male, short in the female. Wing span 40-70 mm.—**Occurrence:** not uncommon in open country during spring. The males fly during the day, the females only at night.—**Diet:** the adults do not feed. The caterpillars (below left) feed on various low plants, including heaths, bramble and sloe.—**Life history and habits:** the caterpillars develop during summer and pupate in an ingeniously constructed pear-shaped cocoon (below right) which has an opening at the front end. The opening has a barrier of silken threads which allows the moth to emerge but prevents the entry of unwanted visitors. The females produce a scent from special glands and the males pick this up with their large antennae, which have an enlarged surface area. They can detect this scent at a distance of several kilometres. As they have reduced mouthparts and do not feed the adult moths only fly for reproductive purposes.

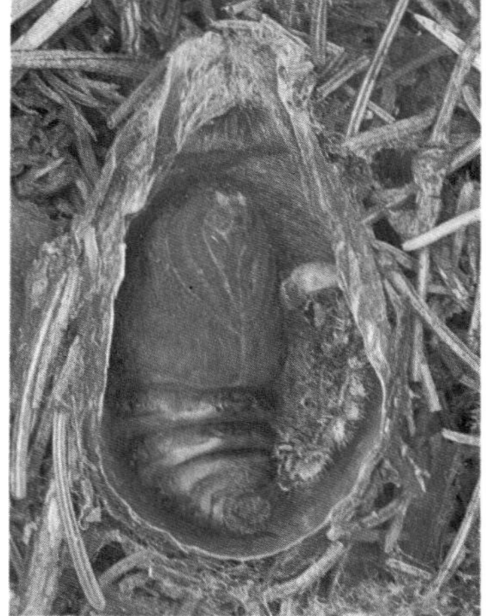

Garden Tiger *Arctia caja* (opposite, above)

Order Lepidoptera (butterflies and moths)

Characteristics: a large moth in which the front wings are whitish with bold brown markings and the hind wings reddish with very dark spots which are often ringed with yellow. Wing span 50-60 mm.—**Occurrence:** widely distributed and fairly common in woods and meadows, flying by night during summer.—**Diet:** the adults do not feed, the caterpillars feed on various plants, including hollyhock and sunflower.—**Life history and habits:** the hairy brown and black caterpillars look like little bears, hence the popular name ''woolly bears''. They are often seen moving across roads. The adults are unpalatable and show warning coloration.

Silver-Y Moth *Autographa gamma* (opposite, below left)

Order Lepidoptera (butterflies and moths)

Characteristics: a small moth with iridescent brownish wings which have a characteristic silvery pattern. The hind wings are paler with a dark border. The antennae are thread-like. Wing span 35 mm.—**Occurrence:** a migratory moth that is quite common during summer and autumn.—**Diet:** the adults suck nectar, the inconspicuous caterpillars live hidden among low-growing wild or cultivated plants and sometimes cause damage to crops.—**Life history and habits:** these are immigrant moths which produce more than one brood in a year, but do not appear to survive the winter in central Europe. When at rest the wings are held over the back like a roof.

Large Yellow Underwing *Triphaena pronuba* (opposite, below right)

Order Lepidoptera (butterflies and moths)

Characteristics: the front wings are grey, brown or yellowish-green with roundish markings, the hind wings bright yellow with a black border. The antennae are thread-like. Wing span 40-60 mm.—**Occurrence:** very common in summer, flying at night, mainly in June and July.—**Diet:** the adults suck nectar, the caterpillars feed on various low plants, including chickweed.—**Life history and habits:** the caterpillars overwinter in the soil, where they pupate in the following spring. One brood per year.

Red Underwing *Catocala nupta*

Order Lepidoptera (butterflies and moths)

Characteristics: a large moth in which the patterned front wings have a bark-like pattern and the hind wings are banded black and red. Antennae thread-like. Wing span 70-80 mm.—**Occurrence:** not uncommon in woodland and among bushes, flying at night in late summer and autumn. One brood in the year.—**Diet:** the adults suck nectar and tree sap, and also rotting fruit. The caterpillars feed at night on poplars and willows.—**Life history and habits:** the eggs overwinter. The large caterpillar, which is whitish-grey mottled with brown, has the shape of a twig, and is therefore very difficult to discern. When resting on tree trunks during the day the moths are well camouflaged, with the inconspicuous front wings hiding the brightly coloured hind wings. When disturbed the front wings are drawn forwards, revealing the hind wings. This frightens the attacker and the moth has time to fly away.

Clifden Nonpareil, *Catocala fraxini*

Order Lepidoptera (butterflies and moths)

Charteristics: a large moth with thread-like antennae. The front wings have a bark-like pattern and the hind wings are dark with a broad blue band. Wing span over 100 mm.—**Occurrence:** not uncommon in woodland, flying at night in late summer and autumn. Not common in Britain.—**Diet:** the adult moths are particularly fond of tree sap and sweet, fermenting substances. They can therefore be baited with a fermenting honey solution. The caterpillars feed on ash, poplar and aspen.—**Life history & habits:** the eggs overwinter and hatch in spring to produce large well-camouflaged caterpillars which are difficult to find.

There are other species of underwing, with red or yellow hind wings. Most of them fly at night and have inconspicuously coloured front wings. They are attracted by the light and therefore enter houses when the windows are open at night. They remain well hidden by day.

Mottled Umber *Erannis defoliaria* (opposite, above)

Order Lepidoptera (butterflies and moths)

Characteristics: umbers are small brownish or grey moths with transverse markings and thread-like antennae. The species shown here are peculiar in that their females have reduced wings; the female (opposite, below left) belongs to the species *Operophtera brumata*. Wing span 40 mm, body length of the female 10 mm.—**Occurrence:** common in woodland, flying at night, in late autumn and often on until the first frosts.—**Diet:** the adult moths do not feed, the caterpillars feed on deciduous trees, frequently on oak, birch and fruit trees, and may cause damage.—**Life history & habits:** in autumn (*O. brumata*) both sexes emerge and the flightless females start to move up the tree trunks where they mate and lay their eggs close to the buds. After overwintering the eggs hatch as the first shoots are appearing. When fully grown the caterpillars, known as loopers, descend to the ground on a silken thread and pupate. On account of the damage they cause fruit growers try to prevent the females from climbing up the trees. A ring of lime wash is often painted round the tree trunks and this stops the females from climbing up to the crown.

There are several other related species, all very similar, but with winged females.

Looper caterpillars (opposite, below right)

Characteristics: These caterpillars have no hairs and only two pairs of legs on the abdomen, in addition to the three pairs at the front of the body. When moving they bend the body in a loop, attach the rear end and then extend the front part, thus moving forwards in a series of loops. They also take up a characteristic position when resting or if threatened. Holding on with only the rear legs the body is raised at an angle and held erect and motionless. In this way it looks like a twig, and this appearance is enhanced by its colour and shape.

White Plume Moth *Pterophorus pentadactyla*

Order Lepidoptera (butterflies and moths)

Characteristics: small moths, in which both pairs of wings are deeply cleft, to form five plumes on each side. This gives a feathery appearance which is enhanced by fringes. The legs are very long, with long spurs. Wing span 20 mm.—**Occurrence:** widely distributed and often seen during summer, when they rest among grasses during the day with the wings rolled up and held out at right angles to the body. They mainly fly at night.—**Diet:** the adults do not feed, the caterpillars feed on various plants, including bindweed.
These very inconspicuous insects belong to a large assemblage of moths known collectively as the Microlepidoptera. These differ considerably in appearance from one another and have varied habits. Some are active at night, others may be active at all times of the day and night, or only in sunshine.

Hautala ruralis

Order Lepidoptera (butterflies and moths)

Characteristics: small moths with thread-like antennae and very broad wings which are usually an inconspicuous grey, brown, or yellowish. When at rest the wings are held extended. Wing span 25 mm.—**Occurrence:** widely distributed among low vegetation in damp places. They fly during the evening and at night and are attracted by the light.—**Diet:** the adults suck nectar, the caterpillars feed on low plants, and pupate in the ground.—**Life history & habits:** there are numerous related species, all very similar to one another in general appearance. As regards their habits, however, there are some very unusual species, including those in which the caterpillars live in water and at first breathe through the skin. Later on these caterpillars live in air-filled cases, which resemble those of the caddisflies.

Cranefly *Tipula* sp.

Order Diptera (true flies)

Characteristics: large insects with a slender body, narrow wings and very long legs. The mouthparts are reduced. The coloration is a very inconspicuous brown. Body length c. 25 mm.—**Occurrence:** widespread and abundant in summer on grassland. They fly at night and remain among the grass during the day.—**Diet:** the adults do not feed. The larvae live in the soil and feed on grass roots and other plant matter.—**Life history & habits:** these delicate flies are often seen in the evening dancing in the air over meadows. These are mostly females which after mating will lay an egg in the soil with their long ovipositor. They are attracted by the light and thus enter houses where they flutter around. Craneflies are quite harmless and cannot sting or bite.

Theobaldia annulata

Order Diptera (true flies)

Characteristics: mosquitoes with a slender body, narrow wings and very long legs. They are inconspicuously coloured. The females have a piercing and blood-sucking proboscis, but the males only have a simple suctorial proboscis. Body length 8-10 mm.—**Occurrence:** widespread and common in the vicinity of water, flying in summer during the morning and evening or when the sky is overcast.—**Diet:** the males only take nectar, but the females bite and suck blood. The larvae feed on small invertebrates.—**Life history & habits:** the larvae are aquatic, but they breathe at the surface. The pupae live in the same way. After emergence the adult females search for a victim from which they can suck blood. Without this blood meal the eggs do not develop. They find the victim by smell and by perceiving the heat emitted by the body of a mammal.

Beech Gall-midge *Mikiola fagi*

Order Diptera (true flies)

Characteristics: if beech leaves are examined carefully it is often possible to see smooth, cone-shaped galls on the leaf veins. Size 5 mm. The tiny flies which have caused these to form are themselves very difficult to find, but they can be reared from the galls.—**Occurrence:** quite common on beech.—**Diet:** the larvae live in the galls and feed on the leaf tissue.—**Life history & habits:** the females lay their eggs in the leaf tissue. The larvae that hatch from these start to produce substances which cause the leaf tissue to proliferate and finally to form a gall (opposite, above left). If a cut is made through one of these galls the almost colourless fly larva will be found inside (opposite, above right) feeding on the plant tissue. The form of the gall and the host plant are always very specific to the insect causing gall formation. See also p. 106 where a gall-wasp is responsible for gall formation.

Bee Fly *Bombylius discolor*

Order Diptera (true flies)

Characteristics: this fly has a thick hairy body and looks rather like a small bumblebee. It has a suctorial proboscis as long as its body. In flight it can hover and remain motionless in the air, and it does this when sucking nectar from a flower. Body length 8 mm.—**Occurrence:** these insects appear in spring, often along the edges of woodland, where in sunny weather they visit flowers. They are widely distributed.—**Diet:** the adults suck nectar, the larvae live as parasites of wasp and bee larvae.—**Life history & habits:** the females lay their eggs camouflaged with sand grains at the entrance to bee and wasp nests. The larvae then enter the nests where they feed at first on the larval food but finally on the larvae of their hosts. Pupation takes place in the closed cells of the host larvae. When the time comes for emergence the pupa uses a spine on its head to break through the cell cap. The fly then emerges.

Hoverfly *Syrphus* sp.

Order Diptera (true flies)

Characteristics: small flies with large, reddish eyes and black and yellow stripes on the body. The wings are very transparent. Body length 10 mm.—**Occurrence:** very common in summer in all sorts of places and often seen hovering.—**Diet:** the adult flies visit flowers, the larvae feed on aphids, sucking out their body contents.—**Life history & habits:** the eggs are laid in the vicinity of aphid colonies where the larvae find plenty of food. Hoverflies, of which there are several species, are therefore very beneficial to man. The black and yellow pattern mimics wasps and this provides the flies with a certain amount of protection. The method of flight is very characteristic, for hoverflies can fly very rapidly and then suddenly come to a stop. When hovering they often remain in the same place for quite a time.

Robber Fly *Machimus* sp.

Order Diptera (true flies)

Characteristics: slender flies with a hairy body and large eyes on the very mobile head. The legs are particularly powerful. When at rest the wings are folded over the back. Body length 15 mm.—**Occurrence:** widely distributed and fairly common along the edges of woodland in summer. They are active when the sun is shining.—**Diet:** the adults are predators which catch other insects. The larvae live as predators among rotting vegetation in the soil.—**Life history & habits:** the adults mostly lurk among vegetation watching out for suitable prey which they fly out and seize. They then return to their resting place with the prey. Robber flies are, however, completely harmless to humans. The eggs are laid on grasses. There are several different species of robber fly, some of which specialise in a certain type of prey.

Grey Flesh Fly *Sarcophaga carniaria* (opposite, above left)

Order Diptera (true flies)

Characteristics: a large insect, somewhat similar to a house fly, but darker and with conspicuous white markings on the body. The eyes are rather large and red. Body length 15 mm.—**Occurrence:** common in the vicinity of houses, indeed often indoors, and seen almost throughout the year.—**Diet:** these and related flies frequent rotting flesh and dung where they feed on the liquids. The larvae feed on flesh and dung.—**Life history & habits:** the larvae develop within the body of the female, who then deposits them on the food. There they liquefy the food source with special ferments and suck up the resulting liquid. Pupation takes place in the soil. On emergence the adult flies work their way to the surface using an inflatable sac on the head to force the end off the pupal case. As flesh flies move about on all kinds of unappetizing substances they pick up bacteria and when they later crawl around on human food they may spread diseases. This is one reason why foodstuffs should always be covered to keep such flies away.

Horse Fly *Haematopota pluvialis* (opposite, above right)

Order Diptera (true flies)

Characteristics: small flies with a grey body and large, iridescent eyes. The mouthparts of the female are modified for piercing. At rest, the wings are folded, roof-like, over the back. Body length 10 mm.—**Occurrence:** the adults appear in summer in damp places and are active particularly on warm, humid days.—**Diet:** the males lick plant saps or nectar, while the females are well known to suck blood. The larvae live as predators, feeding on insect larvae and worms.—**Life history & habits:** the adult females approach their victim silently, fly down and almost immediately pierce the victim's skin with their proboscis. As they do so they inject saliva into the wound and this prevents the blood from clotting, and also causes pain around the wound. This blood meal is necessary for the development of the eggs.

The attractive coloration of the eyes (opposite, below) is seen in many species of horse fly, but it disappears when the insects die.

Sawfly *Allantus* sp.

Order Hymenoptera (ants, bees and wasps)

Characteristics: small wasp-like insects with glassy wings and black and yellow rings round the body. Unlike the true wasps they do not have the characteristic waist joining the thorax and abdomen. Body length 12 mm.—**Occurrence:** widely distributed and not uncommon. They are frequently confused with true wasps. They are active in sunshine, when they visit flowers, especially those of umbellifers.—**Diet:** the adults feed on flower pollen, the larvae are vegetarian.—**Life history & habits:** the females use their ovipositor to pierce slits in stems and leaves, in which they lay their eggs. The larvae resemble butterfly caterpillars, but have a larger number of stumpy legs on the abdomen. There are several species, some causing damage to plants, including gall formation.

Rose Gall-wasp or Bedeguar

Diplolepis rosae (opposite, below left)

Order Hymenoptera (ants, bees and wasps)

Characteristics: it is not uncommon to find large, green and red galls looking like clumps of moss on dog roses. These are known colloquially as robin's pincushions or bedeguar galls and they are caused by the activities of the larvae of this gall-wasp, which live inside them (cf. p. 100).

Cherry Gall-wasp *Cynips quercusfolii* (opposite, below right)

Order Hymenoptera (ants, bees and wasps)

Characteristics: spherical green and red galls on oak leaves are caused by gall-wasp larvae (cf. p. 100). Only females emerge from the galls and these overwinter and then lay eggs in the following spring without mating. The eggs hatch into larvae which produce males and females in the summer. These mate and produce only female offspring.

Digger Wasp of the family Sphecidae

Order Hymenoptera (ants, bees and wasps)

Characteristics: slender, dark insects with brick-red bands on the abdomen. They are frequently seen running about on the ground or among vegetation. Body length about 15 mm.—**Occurrence:** widely distributed and very common, particularly on sandy ground in open country. Active from spring to autumn.—**Diet:** the adults visit flowers, but their larvae feed on butterfly and moth caterpillars previously paralysed by the female.—**Life history & habits:** the female digs a hole in the ground and then goes in search of a caterpillar. When she has found one she injects her venom into it. This paralyses but does not kill the caterpillar, which is then taken to the hole in the ground. The female lays an egg in the caterpillar and closes the hole. The larva feeds on the caterpillar, but without immediately killing it. This does not happen until the final stage of development. The new adult emerges from its hole in the following spring.

Ichneumon of the family Ichneumonidae

Order Hymenoptera (ants, bees and wasps)

Characteristics: slender wasp-like insects with long antennae and legs. The female's ovipositor is also very long. Body length 15-20 mm.—**Occurrence:** ichneumons can be found among vegetation almost everywhere during the summer.—**Diet:** the adults visit flowers, their larvae live as parasites within the bodies of caterpillars.—**Life history & habits:** the females move around among vegetation, almost always with vibrating antennae as they search for a caterpillar. When a suitable host has been found the female uses her fine ovipositor to lay one or more eggs within its body. The larvae at first feed only on those parts of the caterpillar's body which are not essential, e.g. food reserves. It is only at the end of development that the caterpilar is finally killed and the ichneumon larva then pupates.
Ichneumons play a very important part in the control of lepidopteran caterpillars and some species are nowadays bred and used in the biological control of pests.

Ichneumon-fly *Apanteles glomeratus*

Order Hymenoptera (ants, bees and wasps)

Characteristics: small, dark insects of the family Braconidae which are closely related to the ichneumons. Body length 2 mm. The pupal cocoons are very commonly found on dead caterpillars. These are oval, yellow structures, 2 mm in length, and there may be several of them alongside a dead caterpillar.—**Occurrence:** the cocoons are common, particularly in late summer, on house walls or fences in the vicinity of cabbage fields where the caterpillars live.—**Life history & habits:** the adult female ichneumon-fly stings a caterpillar and lays her eggs inside its body. There the larvae, often up to 150 of them, develop until the caterpillar wanders off to pupate when it usually climbs up a wall. The caterpillar then dies and the ichneumon-fly larvae leave it and spin their yellow cocoons alongside it.
These tiny insects are beneficial to man as they help to control pests such as Cabbage White butterflies.

Ruby-tail Wasp *Chrysis trimaculata*

Order Hymenoptera (ants, bees and wasps)

Characteristics: small insects with beautiful iridescent colours, mainly greens, blues and reds. Body length 8 mm.—**Occurrence:** the species illustrated is seen in spring on chalky ground in continental Europe. Several related species are also active during the warmer months in Britain.—**Diet:** the adults visit flowers, the larvae live as parasites in solitary bees and wasps.—**Life history & habits:** the adult females enter the nests of solitary bees and wasps where they lay their eggs. These hatch into larvae which feed on the hosts' food or on their larvae.
The iridescent colours are produced by optical interference and not by pigments, and are similar to those seen in many butterflies.

Wood Ant *Formica rufa*

Order Hymenoptera (ants, bees and wasps)

Characteristics: typical ants, in which the males are completely black, the females and workers black and red. The head carries the kneed antennae and the biting mouthparts. Wings are only present in the males and females but they are cast off after mating. Body length 4-11 mm.—**Occurrence:** widely distributed and common, particularly in conifer forests. This species builds large ant hills, in which the part above ground consists of pine needles and small twigs. These structures may last for several years.—**Diet:** small insects and other invertebrates, as well as sweet juices and aphid secretions.—**Life history & habits:** after mating and casting off her wings, the female or queen starts to establish a new colony. In a suitable place in the soil she builds small chambers in which the first eggs are laid. These hatch and give rise to the first workers (sterile females) which now do all the work of extending the colony, collecting building material, searching for food and tending the larvae and pupae. The conditions within the colony are kept as constant as possible and depending upon the weather the ovate pupae are moved about so that they will develop as rapidly as possible. The worker ants communicate with one another by touching with the antennae, and also when outside the colony by marking with scents. These substances, known as social hormones, also serve in communication within the colony. When large objects have to be transported several ants will co-operate with one another. At certain times winged males and females are produced and these leave the colony to mate during the nuptian flight. A large ant colony may contain over 100,000 individuals. The worker ants move about over a large area, continuously searching for food, and using established pathways, some of which reach the crowns of large trees. Everything suitable is brought to the colony, including many forest pests. Ants can bite and also spray formic acid (opposite, above). The colonies last for several years and they extend deep into the soil, thus providing suitable conditions for overwintering. With the first warm days of spring the worker ants move up towards the surface and start to become active.

Paper Wasp *Polistes gallicus*

Order Hymenoptera (ants, bees and wasps)

Characteristics: these insects are closely related to the Common Wasp and to the Hornet and they have a similar yellow and black livery. They have rather thick antennae and biting mouthparts. The female's ovipositor is modified to form a venomous sting which serves in defence and to capture prey. At rest the wings are folded over the back. These are social insects, living in small colonies. Body length 15-25 mm.—**Occurrence:** usually common and widely distributed in summer and autumn in Europe, but not in Britain which has several other wasp species.—**Diet:** the adults feed mainly on nectar and fruit, and they feed their larvae on masticated insects.—**Life history & habits:** a wasp colony has a queen, a number of workers and some males. They live in a nest which is constructed out of a type of paper made by masticating wood pulp. Only the mated queens survive over the winter. In spring each queen starts to build a nest and lay eggs from which she rears the first workers. She then lives only to lay eggs, leaving all other tasks to the workers. They catch the prey, chew it up and feed it to the larvae. The young queens and males leave the nest in autumn to mate. The workers no longer have duties in the nest and they are often seen at this time feeding on the juices of windfall apples and other fruits. After mating the queens go off to find a place to spend the winter. All the workers and males die in the autumn.

There are several wasp species in Europe, including the very large Hornet. They differ in their habits and nest structure. Some suspend the nest from a branch, while others build nests in hollow trees or in holes in the ground. When irritated wasps use the sting to inflict painful wounds. Bees have a barbed sting which cannot be withdrawn from the victim but is torn from the insect's body when it flies off. Wasps, on the other hand, have a smooth sting which can be withdrawn. A wasp can therefore sting several times.

Common Carder Bee *Bombus agrorum*

Order Hymenoptera (ants, bees and wasps)

Characteristics: a large bumblebee with a hairy body, bright coloration and sucking mouthparts. The hind legs have special baskets for the collection of pollen; these are formed by the hairy tibiae. Body length 12-20 mm.—**Occurrence:** found almost everywhere. The species illustrated here is commonly seen in open country, fields and meadows, often visiting flowers, and it flies from spring until autumn.—**Diet:** both adults and larvae feed on flower pollen and nectar, the latter partly in the form of stored honey.—**Life history & habits:** as in the wasps, it is only the mated females that overwinter. In the spring each female seeks a hole in the ground, often a mouse hole, and begins to make a nest (opposite, below). The base consists of grass stalks, moss and mouse hairs. On this the female or queen places a mixture of nectar and pollen, known as bee-bread, on which she lays her first eggs, covering them over with wax to form a cell. The eggs hatch into young larvae which feed on the bee-bread and later on extra food brought to the nest by the queen. When ready to pupate each larva spins a cocoon within its cell. Later, when the new bees have emerged these cells are used for storing honey. These first generation bees are workers which now help to rear subsequent generations. They visit flowers and are of great importance as pollinators of many plants. They have a much longer proboscis than a honey-bee and can therefore reach to the bottom of long tubular flowers, such as red clover. In the absence of bumblebees these flowers are not pollinated. Bumblebee colonies are never very large, most having only a few hundred inmates. In the late summer the colony produces males and females. The males hatch from unfertilized eggs. The males and females then leave the colony on what is known as the nuptial flight, during which mating takes place. The mated females seek a place in which to overwinter. All the males and workers die in the autumn. Bumblebees are friendly insects which do not attack humans, but if picked up they can inflict a painful sting.

Honey-bee *Apis mellifera*

Order Hymenoptera (ants, bees and wasps)

Characteristics: the well-known honey-bee has a dark brown, slightly hairy body, and glassy wings with a pale brownish tint. The hind legs of the workers have pollen baskets for the collection of pollen, an ovipositor in the form of a sting and mouthparts that are modified for sucking. The males or drones (opposite, below, centre left), which have no sting, are larger and broader than the workers and their large eyes meet in the middle of the forehead. The queen resembles a worker, but is considerably larger. Body length 15-21 mm.—**Occurrence:** widely distributed as one of man's domesticated animals. Wild or feral bees are rather rare. Honey-bees fly in all the warmer months and their colonies last for several years.—**Diet:** the adults and larvae feed on pollen and honey or nectar.—**Life history & habits:** honey-bees live in large colonies or hives, consisting of vertical combs of regular hexagonal cells. Here the queen lays innumerable eggs during the course of a life which may last several years. The larvae are tended and fed by the young workers. Older workers are employed in collecting food. They visit flowers and on their return to the hive perform a special dance which tells the other bees the distance and direction of good nectar sources. The nectar is stored as honey in special combs. Larvae destined to develop into queens are fed on "royal jelly", a salivary gland secretion produced by the workers that is particularly rich in protein. These larvae live in larger (queen) cells and develop into fertile females, whereas the larvae in smaller cells fed on pollen and nectar only become workers (infertile females). The drones develop from unfertilized eggs. After the emergence of a young queen, the old queen leaves the hive together with a swarm of workers. The first young queen then kills all the larvae living in queen cells, and leaves the hive with the drones for a nuptial flight during which she mates with several drones. She then returns to the hive. A honey-bee colony may have a population of more than 50,000. In autumn when the colony is preparing for the winter, all the drones are killed.

Green Tiger Beetle *Cicindela campestris*

Order Coleoptera (beetles)

Characteristics: iridescent green beetles with a few yellow markings on the elytra and relatively long legs and antennae. The mandibles are long and pointed. Body length 12-16 mm.—**Occurrence:** widely distributed and quite common in open woodland and other sunny places from April to June.—**Diet:** both adults and larvae prey on other insects.—**Life history & habits:** tiger beetles are sun-loving insects which live mainly on the ground where they hunt their prey. They run fast and when disturbed they can fly very well, but never for long distances. Although brightly coloured they remain well camouflaged on the ground. The prey is seized with the powerful mandibles and pre-digested by digestive juices, the resulting liquid being then sucked up by the beetle. Larval development takes two years and thus adults and larvae can be found at the same time. The larvae form tunnels in the ground within which they lie in wait for prey which is quickly seized and consumed in the same way as the adults. They pupate in a hole in the ground. When the adult beetles emerge both pairs of wings are still small and soft, but they become extended to their full size by the inflow of blood under pressure. They then harden in the air, the front wings forming the elytra, beneath which are folded the transparent hind wings. Only the hind wings are used in flight. Thus, most beetles can fly, even when they are very large, but there are a few in which the hind wings are reduced. There are eight species of tiger beetle in Europe, of which four, including the species illustrated, occur in Britain.

Carabus auratus

Order Coleoptera (beetles)

Characteristics: a large, slender iridescent green and golden ground beetle with longitudinal furrows on the elytra. The legs and mandibles are powerful and the antennae long. Body length 20-27 mm.—**Occurrence:** widely distributed and abundant, particularly in fields and meadows during spring and summer.—**Diet:** both the adults and larvae are predatory, taking worms, snails and small insects.—**Life history & habits:** these beetles remain hidden by day, but come out at night to hunt. Prey is seized and torn to pieces by the mouthparts. If attacked the beetle sprays an evil-smelling liquid from glands near the vent, and this repels intruders. The larvae also move around in search of prey.
There are several other ground beetles, all similar in size and appearance, and with similar habits.

Pterostichus nigra

Order Coleoptera (beetles)

Characteristics: a smaller iridescent black ground beetle with numerous longitudinal ridges on the elytra. Body length 15 mm.—**Occurrence:** this species is very common in grassy places, and can be seen throughout almost the whole year. It is also active by day.—**Diet:** insects and other small invertebrates.—**Life history & habits:** the beetles overwinter and lay their eggs in spring. The larvae develop during the summer, pupate and emerge as adult beetles in the autumn. As these live for a long time they can be seen during many months of the year.
There are several other species in this genus, most of which are smaller. They are usually iridescent black.

Diving Beetle *Dytiscus marginalis*

Order Coleoptera (beetles)

Characteristics: a large flattish beetle with an oval outline. The hind legs are broad and flattened with swimming hairs. This species is dark olive-green with a yellow border to the body. The males have smooth elytra and very small suction pads on the front legs, the females have ridged elytra and normal legs. Body length: 30-35 mm.—**Occurrence:** found throughout the year in quiet waters, where they swim about underwater.—**Diet:** both larvae and adults are voracious predators which attack prey of any suitable size, including tadpoles and fishes.—**Life history & habits:** the larvae are also aquatic. They seize prey with the mandibles which inject digestive juices from the alimentary canal into the prey. These juices pre-digest the tissues of the prey and the insect then sucks up the resulting fluid. They pupate in a hole above the water line. The adult beetles feed in the same way, but have to come up to the surface to renew their air supply, which is held by the fine hairs at the end of the abdomen. They can fly well and in this way colonize new areas of water.

Staphylinus caesareus

Order Coleoptera (beetles)

Characteristics: small rove beetles with a narrow body and very short elytra which only partly cover the abdomen. In spite of this the hind wings are fully developed and the beetles fly well. The thorax is dull black, the elytra red-brown. There are iridescent spots on the sides of the body. Body length 25 mm.—**Occurrence:** widely distributed, mainly among rotting vegetation on the ground, where they run about rapidly.—**Diet:** probably insects and their larvae living in the rotting vegetation.—**Life history & habits:** there are several species of rove or staphylinid beetles, mostly small, and it is characteristic that the abdomen is curled upwards when they are moving. Some species live in ant nests, where they secrete sweet substances which are licked up by the ants. They feed on waste matter and also on the ant larvae.

Dor Beetle *Geotrupes* sp.

Order Coleoptera (beetles)

Characteristics: dor or dung beetles are broad and rounded iridescent black insects with ridged elytra. Their legs are short and powerful, the tibia of the front legs having powerful spines which help in digging. The short antennae have a series of densely packed lamellae at the end. Body length 20-25 mm.—**Occurrence:** found during all the warm months in open places or along the edges of roads, particularly where there is animal dung. They are mainly active at twilight.—**Diet:** the adults dig holes in the ground which they fill with dung on which the larvae feed.—**Life history & habits:** both sexes usually take part in burying the dung. They dig more dung into the ground than is used by the larvae and thus accelerate its decomposition.

Stag Beetle *Lucanus cervus*

Order Coleoptera (beetles)

Characteristics: very large beetles with a broad head and long legs. The antennae have lamellae at the ends. The head and thorax are black, the elytra dark brown. In the males the mandibles are modified to form large antler-like structures; those of the females are considerably shorter. Body length 30-80 mm.—**Occurrence:** found in summer, mainly in oak woods, but becoming scarce in some areas.—**Diet:** the adults suck the sap exuding from the trees, the larvae live on the wood of old, hollow oaks.—**Life history & habits:** the larvae take 4-6 years to complete their development, and then pupate. The adults are active at night, when they fly around. The males fight over the females and continue to do so until one lands on its back and gives up.

Rose Chafer *Cetonia aurata*

Order Coleoptera (beetles)

Characteristics: very large iridescent green beetles with white flecks on the elytra and coppery-red undersides. The antennae have lamellae at the ends. Body length 15-20 mm.—**Occurrence:** not uncommon in summer, on flowers or flying around flowering shrubs, but only active in sunlight.—**Diet:** The adults feed on the petals of flowers, particularly dog roses. The larvae feed on rotting vegetation in tree stumps and among roots.—**Life history & habits:** Rose Chafers fly with the elytra closed over the body. The eggs are laid in suitable places and the larvae feed for several years on rotting vegetation. This is a poor source of nourishment hence they take a long time to develop.
Related chafers in the tropics are among the most beautiful insects, showing brilliant iridescent colours, which are physical in nature and comparable to those seen in butterflies.

Anomala dubia

Order Coleoptera (beetles)

Characteristics: this is a brown scarab beetle with a slightly hairy upperside. The antennae have lamellae at the ends. Body length 15-20 mm.—**Occurrence:** very common in summer in bushes and trees in gardens and woodland.—**Diet:** the adults feed on leaves, the larvae on the roots of grasses and other plants.—**Life history & habits:** the subterranean larvae move about very little. They take two years to become fully grown and then pupate in the ground. The adults are often seen in sandy places, and usually near the coast. When present in large numbers they consume large amounts of foliage. Sometimes the larvae also cause damage as they eat roots. There are several similar species.

Cockchafer *Melolontha melolontha*

Order Coleoptera (beetles)

Characteristics: a well-known fairly large beetle with brown, coarsely ridged elytra, a black body with white triangular markings on the flanks and a black thorax. The legs and antennae are brown, the latter with a group of lamellae at the tips (opposite, above right); each antenna has 7 of these lamellae in the male, 6 in the female. The powerful legs have spines and large claws. The females can also be distinguished by the conical ovipositor at the rear end of the body. Body length 20-30 mm.—**Occurrence:** in deciduous forests and along the edges of woodland, and very abundant in the spring of some years, particularly in warm localities. The adults fly mainly in the evening and are not normally active during the day.—**Diet:** the adults feed on the leaves of deciduous trees, particularly beech and oak, often causing considerable damage. The larvae live in the soil, feeding on the roots of various plants, and also causing damage when present in large numbers.—**Life history & habits:** the larvae (opposite, above left) take three years to complete their development. After their emergence from the pupa the adults move to the trees where they immediately start to feed on leaves. After mating the female digs quite deeply into the soil and lays about 70 eggs. The larvae hatch and soon begin to feed on roots. In the summer of the third year they pupate in a hole in the ground and emerge as adults in the autumn of the same year. They remain, however, in the ground until the following spring and then leave their hole. Various animals, such as foxes and falcons, prey on these beetles.
There are other related species, one of which, *Melolontha hippocastani*, occurs locally in Britain. Its larvae take 4-5 years to develop, and they cause damage to the roots of young conifers in plantations.

Glow-worm *Lampyris noctiluca*

Order Coleoptera (beetles)

Characteristics: a small beetle in which the female (illustrated) is wingless, but the male is winged. The special characteristic is that both adults and larvae produce light. The adults are an inconspicuous grey-brown with short antennae. Body length 10-18 mm.—**Occurrence:** widely distributed in damp places on calcareous soils where their food occurs. Seen from June to August.—**Diet:** the adults scarcely feed at all, but the larvae prey on snails and slugs.—**Life history & habits:** the larvae hunt on the ground. The adults are active at night and the females attracts a mate with her luminescent organ which is on the abdomen. The light is greenish and produced by a chemical reaction. It can be switched on and off, and it does not produce heat.

Click Beetle *Elater* sp.

Order Coleoptera (beetles)

Characteristics: small beetles with a narrow body, long antennae and usually ridged elytra which are pointed posteriorly. They have a special spring mechanism which enables them to turn themselves over when lying on their backs. Body length 10-20 mm.—**Occurrence:** very common among vegetation in summer.—**Diet:** the adults feed on flowers and plant sap, but also eat other parts of plants. The larvae feed on seeds and plants and may cause damage.—**Life history & habits:** the larvae live in the soil and take scarcely a year for their development. The spring mechanism of the adults consists of a spine on the underside of the prothorax which catches against the edge of a pit on the mesothorax. When the beetle is lying on its back the spine slips out of position and the elytra strike the ground so that the beetle is jerked into the air, to land on its feet. They also fly in the normal way.

7-spot Ladybird *Coccinella septempunctata* (opposite, above left)

Order Coleoptera (beetles)

Characteristics: a small round ladybird with a much arched back and red elytra marked with black dots. The thorax is black and pale yellow. Body length 6-8 mm.—**Occurrence:** widely distributed and common throughout almost the whole year, particularly in places with numerous aphids. Some adults overwinter in buildings.—**Diet:** adults and larvae feed principally on aphids.—**Life history & habits:** the larvae live out in the open among aphid colonies and help to control these pests. The adults overwinter under stones, in old walls or in crevices in bark.

Anatis ocellata (opposite, above right)

Order Coleoptera (beetles)

Characteristics: a ladybird in which the numerous black dots on the elytra are each encircled by a pale ring. Body length 10 mm.—**Occurrence:** not so abundant and widespread as the preceding species.—**Life history & habits:** this species also feeds on aphids. The bright colours serve as a warning that the ladybird is unpalatable.

Colorado Beetle *Leptinotarsa decemlineata*

Order Coleoptera (beetles)

Characteristics: a small roundish beetle with a high-arched back and relatively short antennae. The legs are not particularly powerful and the beetles only walk slowly, but they fly well. The elytra have striking black and yellow stripes.—**Occurrence:** originally native to North America, but now a serious pest in many parts of Europe. Fortunately not established in Britain, although a few individuals are occasionally found in the south-eastern counties.—**Diet:** adults and larvae feed on potato plants.—**Life history & habits:** the adults overwinter and the larvae are found in early summer on potato plants, often in great numbers and causing very serious damage. When threatened both larvae and adults produce droplets of a coloured fluid which has a defensive function.

Chrysomela sp.

Order Coleoptera (beetles)

Characteristics: small roundish leaf beetles with green or blue iridescence. They do not move about much and mostly live on plants where they hold on with the legs. Body length 8-10 mm.—**Occurrence:** abundant among vegetation during the summer, but usually only on their own special food plant, which varies according to the species.—**Diet:** vegetation.—**Life history & habits:** the larvae and adults feed on a variety of plants, usually cutting holes in leaves, and some are very serious pests. The beautiful iridescent colours of these beetles are due to the physical structure of the elytra and other parts of the exo-skeleton. There are several species of leaf beetle which are often difficult to distinguish. Some species may be identified by the plant on which they are feeding, as in many cases this is quite specific.

Saperda carcharias

Order Coleoptera (beetles)

Characteristics: a large but narrow beetle with powerful legs and very long antennae (hence the popular name of longhorn beetles for the family). The general coloration is brownish and the elytra have numerous very small black pits and dense yellowish hairs. Body length 20-28 mm.—**Occurrence:** the adults are found in summer in the vicinity of the larval food plants.—**Diet:** the larvae live in the trunks and roots of various poplars and their long tunnels in the timber may cause serious damage. The adults feed on leaves and tree sap.—**Life history & habits:** larval development takes from two to four years. They then pupate beneath the bark of the tree and later emerge as adults. The latter are often seen resting on poplar trunks, and they also fly well.

Some of the tropical longhorn beetles are among the largest of all insects.

Strangalia maculata

Order Coleoptera (beetles)

Characteristics: a typical longhorn beetle with a narrow body and long antennae. The elytra are yellow with black transverse bars, showing a slight resemblance to wasps. As the adult beetles visit flowers this wasp-like pattern may help to protect them from predators. Body length 17 mm.—**Occurrence:** not uncommon in summer along the edges of woodland and in forest clearings, particularly on umbelliferous flowers.—**Diet:** the adults feed on parts of the flowers, the larvae live in the fallen trunks of deciduous trees, especially in the rotting timber.—**Life history & habits:** the adults are active by day, particularly when the sun is shining. In addition to their bright coloration, they are also striking because they produce high-pitched sounds. It is not clear whether these are used as a warning or as a signal to a potential mate. The latter is more probable as insects can perceive high notes.

Leptura rubra

Order Coleoptera (beetles)

Characteristics: another typical longhorn beetle with a narrow body and long antennae. The elytra are brilliant red. Body length 18 mm.—**Occurrence:** very common during summer in Europe, along the edges of woodland where the adults often visit flowers, especially those of umbellifers. In Britain recorded locally in Norfolk. Diet: the adults feed on parts of the flowers. The larvae live in the rotting wood of fallen conifers, and sometimes in telephone poles.—**Life history & habits:** the diet of rotting wood is not very nutritious and so it is not surprising that the larvae take several years to complete their development.

There are many other species of *Leptura* which can be seen visiting flowers alongside flies and dragonflies. In some species the antennae are longer than the body.

Otiorrhynchus sp.

Order Coleoptera (beetles)

Characteristics: this is one of the weevils, a family of beetles containing over 40,000 species, in fact it is the largest family in the animal kingdom. The characteristic feature is the protruding snout or rostrum carrying the mouthparts at the tip, and the normally elbowed antennae. Weevils are mostly small and roundish with an arched back. In most species the coloration is dull and inconspicuous. Body length 10-15 mm.—**Occurrence:** the weevil illustrated here comes from mountain forests in central Europe, where it lives on conifers and is not uncommon in summer.—**Diet:** like all weevils both the adults and the larvae are vegetarian, the adults gnawing young conifer shoots and thus causing damage.—**Life history & habits:** the larvae are legless grubs living in plants stems and roots. In this genus reproduction may be parthenogenetic, that is, the female may lay eggs which hatch without fertilization.

Cantharis sp.

Order Coleoptera (beetles)

Characteristics: small soldier beetles with a narrow body and long antennae. The elytra and body are very soft, a character which distinguishes soldier beetles from the rather similar longhorns. The elytra are dark, the rest of the body brownish. Body length 10-15 mm.—**Occurrence:** widespread and common in summer, flying by day and visiting flowers.—**Diet:** the adults and larvae prey on other insects.—**Life history & habits:** the larvae overwinter and can be seen hunting on the ground in early spring. The larvae seize the prey with their hollow mandibles and through them inject a secretion from their gut which pre-digests the prey's tissues. The larva then sucks up the resulting fluid. The adults are seen later in the year, often on the flowers of umbellifers.
The family Cantharidae contains several related beetles, with similar habits.

ILLUSTRATED INSECTS

English Names

Scientific Names